Civil PE Structural Practice Exams: 40 Civil and 80 Structural Practice Problems with Quick Reference Guide

Study More Efficiently

David Gruttadauria P.E.
PECoreConcepts@gmail.com

Thank you for your Purchase!

We are available for full support of your studying needs and any questions or comments you may have. Email us at:

PECoreConcepts@gmail.com

For **<u>40 Free Additional Civil Questions</u>**, go to the Answer Key page to see how to subscribe your book.

Table of contents

C. Traffic Volume
D. Vehicle Dynamics

7. Materials
    A. Soils Classification
    B. Soil Properties
    C. Concrete Properties
    D. Steel Properties
    E. Material Test Methods and Specification Conformance
    F. Compaction
8. Site Development
    A. Cut and Fill
    B. Construction Site Layout and Control
    C. Construction Erosion and Sediment Control
    D. Impact of Construction on Adjacent Facilities
    E. Safety

<u>Structural Depth</u>

9. American Society of Civil Engineers (ASCE)
    a. Live Loads
    b. Snow Loads
    c. Snow Drifts
    d. Occupancy and Site Class
    e. Site Coefficients and Spectural Response
    f. Effective Seismic Weight
    g. Seismic Base Shear and Distribution of Forces
10. AASHTO
    a. Introduction to Bridges
    b. Limit States and Load Factors
    c. Live Load Distribution
11. American Concrete Institute (ACI)
    a. Flexure
    b. Shear
    c. 2-Way Shear
    d. Axial
    e. Reinforcing Development and Details
12. American Institute of Steel Construction (AISC)
    a. Flexure
    b. Compression
    c. Tension
    d. Fatigue
    e. Bolt Strength
    f. Block Shear
    g. Weld Strength

13. National Design Standards for Wood Construction (NDS)
    a. Sawn Lumber
    b. Structural Glued Lumber
    c. Connections
14. ACI 530 Specifications for Masonry Structures (MSJC)
    a. Bending
    b. Compression
15. Precast/Prestressed Concrete Institute (PCI)
    a. Prestressing Concepts
    b. Prestressing Stresses
    c. Flexure
    d. Shipping and Handling
16. OSHA
    a. Safety Concepts
17. IBC
    a. Loads
    b. Fire Rating
    c. Special Inspections
18. AWS
    a. Welding Symbol
    b. Basic Weld Types
19. Advanced Statics
    a. 3D Loads
    b. Moving Loads
    c. Hinges
    d. Cables
20. Miscellaneous
    a. Eccentrically Loaded Bolt Groups
    b. Elongation/Contraction Due to Axial Loading
    c. Thermal Expansion/Contraction
    d. Composite Members
    e. Torsional Shear
    f. Surcharge
    g. Deflections
    h. Horizontal Shear Stress

II. Practice Problems

III. Solutions

IV. Answer Key

CORE CONCEPTS QUICK REFERENCE GUIDE

MORNING BREADTH

Project Planning

Quantity Take-off Methods

Quantity take-off methods are a means for estimating the cost of each aspect of a project. A project consists of many activities and materials all of which are accounted for as items of a project. For example, a project may involve the construction of a retaining wall. There are many activities and materials associated to complete this. Some include excavation, formwork, concrete for the wall, reinforcing steel etc. When contract drawings and specifications are developed, all of these items must be identified. All items also must include a quantity associated it's them to indicate the amount or extent of work for the item. These quantities must be defined by a particular unit of measure which must be appropriate for the action or material. Taking excavation as an example of an item, there must be an amount of excavation associated with it. Since excavation involves removing a volume of material, the most appropriate unit is cubic yard or cubic feet. To estimate the cost of the project, each item has a price per unit associated with it. This price is determined by previous similar work and taking into account the specifics of the particular project. Below is an example of the breakdown of some items associated with an example retaining wall project:

Item	Unit	Quantity	Unit Price	Cost of Item
Excavation	Cu. Yard	50	100	5000
Concrete (Including Formwork and labor)	Cu. Yard	25	1000	25000
Reinforcing Steel	Lb.	500	12	6000
Backfill	Cu. Yard	40	50	2000
Drainage Pipe	Linear Ft	30	5	150

Cost Estimating

Engineering Economics is used to determine the best economic course of action when weighing construction options by incorporating the life-span of alternatives and comparing costs at equivalent times. The following chart provides the most common equations;

Converts	Formula
P to F	$(1+i)^n$
F to P	$(1+i)^{-n}$
F to A	$\frac{i}{(1+i)^n - 1}$
P to A	$\frac{i(1+i)^n}{(1+i)^n - 1}$
A to F	$\frac{(1+i)^n - 1}{i}$
A to P	$\frac{(1+i)^n - 1}{i(1+i)^n}$

P = Present Value
F = Future Value
i = Interest Rate
n = Years
A = Uniform Series Value

Project Schedules

Project schedules must be set and maintained to ensure it remains on time and on budget. To determine a project schedule, all tasks must be identified and the length of time (durations) for each task must be estimated. These tasks can then be sequenced by determining what the appropriate order of tasks are. Some tasks must be completed before others can begin. These tasks are defined as predecessors. See the example chart below indicating identified tasks, durations, and predecessors:

Task	Duration (Days)	Predecessor
A	2	
B	3	A
C	2	A
D	1	B
E	2	B, C

This information can then be visualized by producing and activity diagram. First begin by drawing tasks. Start with A:

A

Then determine which tasks have A as a predecessor. Draw these tasks as well with arrows indicating these tasks are connected:

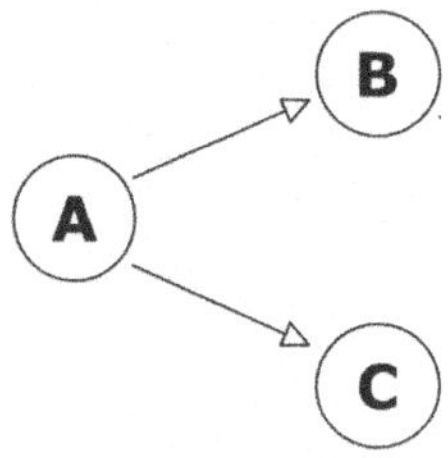
B
A
C

Continue in the same manor with each task. The final chart is as follows:

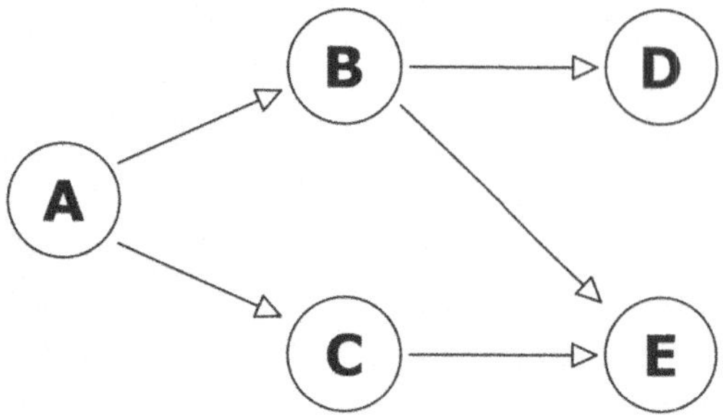
B
D
A
C
E

Then you can determine the critical path of the project. The critical path is defined as the sequence of tasks which would yield the shortest amount of time to complete the project. If the duration of any task on the critical path is changed, the duration of the entire project will change. In the example above, you can determine the critical path by identifying all paths and the critical one is the longest sum of duration. Therefore, the possible paths are A-B-D, A-B-E, and A-C-E which have total durations of 6, 7, and 6. Therefore the critical path is A-B-E. A change in duration of non-critical tasks will only change the project duration if the change creates a longer path than the existing critical one.

Activity Identification and Sequencing

The appropriate steps in the proper sequence need to be identified to complete a project. This involves understanding all the tasks involved in a specific project type and providing a timeline of events to properly facilitate the successful completion of the project. There are many types of projects and the specifics can vary. For the purposes of the PE Exam, it is important to have a general knowledge of common construction tasks and sequences. Below are some examples of design and construction tasks divided by when they occur in certain project phases:

Pre-Design/Design/Project Award

- Owner initiates project
- Owner hires Architect/Engineer or uses In-House Architect/Engineer
- Contract documents and specifications are developed
- Contractors bid on the project
- Project is awarded

Pre-Construction

- Contractor submittals are reviewed and approved
- Sub-Contractors hired
- Site survey, staking, and layout
- Procurement of materials

Construction

- Traffic Control, water handling, etc. installed if necessary
- Crane set up and positioning
- Temporary earth retaining systems installed if necessary
- Excavation
- Formwork or Erection
- Testing of materials
- Installation of rebar
- Pouring of concrete
- Concrete curing
- Backfill

Post-Construction

- Semi Final/Final Inspections
- Open road to traffic
- Punch-Lists
- As-Built drawings

Means and Methods

Construction Loads

Construction loads are temporary loads, occurring within the duration of a project, imposed on a structure which may be partially or fully complete. This may include materials, personnel, equipment, or temporary structures. The concern for construction loads is to understand the different types of stresses they may impose on members as opposed to the final in-place condition of those members and ensure they are designed to handle these forces.

Materials: Storage of materials is an often overlooked aspect of a project. Rebar, excavated materials, or other building materials need to be stored in an accessible location and will often impose a large additional dead load on the structure.

Temporary Structures: Temporary structures may often be needed to either provide additional support to unstable members or access for personnel to continue the erection process. Temporary structures may also be for the housing of materials or personnel.

Equipment: Equipment is often needed for various construction activities such as welding or painting procedures. The weight and distribution of these loads should be accounted for.

Cranes: Cranes can also be considered equipment however special attention should be given to the sequencing of erection based on the cranes reach.

Members in Temporary Conditions: Along with additional dead load, construction can introduce stresses into members for which they are not designed. Some examples include the erection of a precast member such as a wall panel which may be designed for compression but will see some flexure about its weak axis while it is being picked and placed. Also, the first steel girder in a bridge before it is connected to the others through diaphragms will be unstable and must be temporarily supported. In these conditions, design measures need to be taken even though they are not required for the final condition.

Construction Methods

Steel: Strong and durable material. Steel has the capabilities to be used for long span bridges and high-rise buildings. Steel members are manufactured using either the hot-rolled or cold-formed methods. Steel members are provided in predetermined shapes. Some examples include W-, S-, C, and HSS-shapes. Steel is connected and constructed by the use of bolted or welded connections. The advantages of steel are again the ability to span long distances and the weight of the members compared to the strength is relatively low. Some disadvantages include the high cost, lack of ability to form unique shapes, and tendency of the material to corrode. When steel is exposed to salts, a chemical reaction occurs causing the steel to rust and even loose section properties. To counteract corrosion some preventive measures are paint systems, coating systems such as galvanizing, or weathering steel.

Reinforced Concrete: Concrete is strong in compression but weak in tension. However it has the ability to bond to reinforcing steel to appropriately resist tension. Reinforced concrete is used in buildings and shorter span bridges or certain components of bridges. Some common applications are foundation elements, bridge decks and parapets, or retaining walls. The advantage of concrete is it can be formed to any shape or aesthetic look with proper formwork and is strong in compression. The disadvantages however are that concrete has a high self-weight, will likely crack, and has a limited span length. The reinforcing in concrete can also corrode and cause pop-outs or spalls.

Precast/Prestressed/Post-Tensioned Concrete: Precast concrete is concrete which is cast somewhere other than its final location, either at a plant or another area on the construction site, and is then stripped from its forming, transported to the site, and erected. Prestressed concrete is precast concrete which has been pre-compressed using steel strands with high elasticity. The strands are tensioned to a design force before the concrete is cast. Then the concrete mix is placed and cured. The strands are then cut at the ends. Since the strands have a high elasticity, they will try to return to their original state. However, since the stands are now bonded to the concrete, there is a compressive force transferred to the concrete. This force will oppose the stresses caused by bending. Post-Tensioned concrete uses the same concept as prestressed concrete. However, the concrete member is cast first and the strands are tensioned through the member using plastic tubes embedded along the length of the member. The tube is then grouted and the strands are cut to transfer the force. Precast Concrete will have the lowest tensile capacity and therefore is used for the lowest spans. Precast is often used for compression members or bridge deck units. Prestressed Concrete is be able to span larger distances and is used for floor members. It is common in parking garages as double-tee shapes for floor members or for long span bridges with common shapes such as prestressed bulb tees. Post tensioned concrete is not as common and is used for much larger spans. Precast concrete advantages are the quality of concrete is often better under plant controlled conditions and the construction is much quicker. The disadvantages are there is a higher cost than reinforced concrete due to shipping and erection expenses and the tendons also are susceptible to corrosion.

Wood: Relatively low strength material. Wood is often used in residential applications or for very small span bridges. Wood is extremely cheap and lightweight for erection. In addition to the low strength, wood also will deteriorate due to rot and is highly sensitive to fire damage.

Masonry: Can be reinforced or unreinforced. Masonry is also strong in compression but weak in tension. Only used in small retaining wall applications and some older bridges still are composed of masonry components.

Temporary Structures

Structures which are built for a specific purpose, often to facilitate an aspect of a construction project, and are removed before the conclusion of the project are temporary structures. Some examples include:

- Temporary Buildings: May be used for storage or offices during construction.
- Scaffolding: Temporary elevated platforms which provide access to perform certain tasks.
- Temporary Supports and Shoring: Often forces are introduced during construction which are not the same as the final in-place conditions. In these situations, temporary supports are needed to keep structural members stable until the construction can be complete.
- Temporary earth retaining systems: When excavation is needed, there is often not enough room to safely dig to the required depths. There may be a need to support traffic or adjacent facilities during the excavation. In these cases, temporary earth retaining is required. Some examples include sheet piles, concrete blocks, or trench boxes.
- Formwork: Concrete formwork is used to place concrete to the desired shape and will remain in place until the mix has cured to the desired strength.
- Cofferdams: A wall constructed to prevent the flow of water to a specific area. Can be made of sandbags, sheet piling, or other materials.

Soil Mechanics

Lateral Earth Pressure

Rankine Active Earth Pressure

Resultant Force $R_a = ½ k_a\gamma H^2$ which is applied at a distance of H/3 from the base of the footing

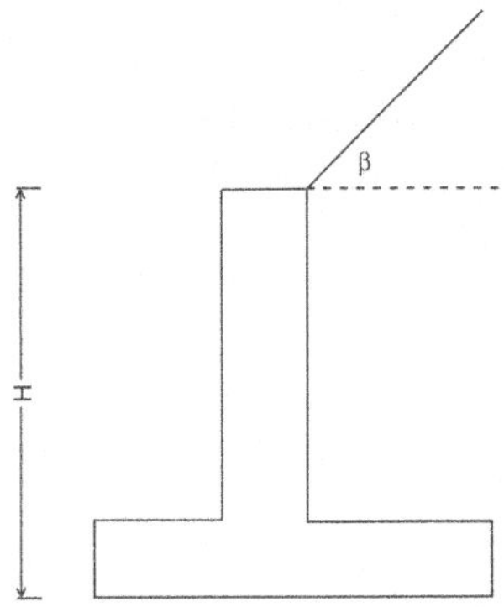

β
H

k_a = Active Earth Pressure Coefficient
γ = Density of Soil
H = Height of Retaining Wall from Base of Footing

$$k_a = \cos\beta \left(\frac{\cos\beta - \sqrt{\cos^2\beta - \cos^2\phi}}{\cos\beta + \sqrt{\cos^2\beta - \cos^2\phi}}\right)$$

β = Angle of Backfill
ϕ = Angle of Internal Friction
If the backfill is horizontal ($\beta = 0$) the equation reduces to:
$k_a = \tan^2(45^\circ - \phi/2)$

Note: Rankine assumes the friction between the soil and wall is zero

Soil Consolidation/Foundation Settlement

Settlement is when the soil supporting the foundation consolidates which causes a decrease in volume and a drop in elevation. This causes the foundation to no longer be fully supported and will introduce additional stress. There are three phases of settlement:

1. Immediate Settling or Elastic Settling: This settlement occurs immediately after the structure is built. The load from the structure causes instant consolidation of the soil. This is the main component of settlement in sandy soil conditions.

2. Primary Consolidation: A more gradual consolidation which is due to water leaving the voids over time. This is mostly a factor only in clayey soils.

3. Secondary Consolidation: Also occurs at a very gradual rate. This is due to the shifting and readjustment of soil grains. Most often this is the lowest magnitude of consolidation phases.

Effective Stress

The effective stress is the stress at a certain point below grade due to the weight of soil above. It is calculated as the density times the height of each level. However, If the water table is present, the density is reduced by that of water:

Effective Stress = $\Sigma H\gamma$ in dry soil and $\Sigma H(\gamma_s - \gamma_w)$ in saturated soils, where γ_w = 62.4 pcf

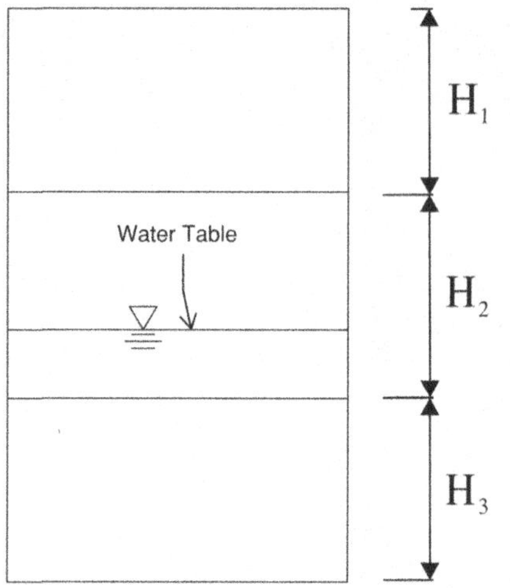
Water Table
H_1
H_2
H_3

Bearing Capacity

For shallow foundations, the soil below must be suitable to support the load transferred through the footing. Different types of soils have different bearing capacities. Sand is often a good foundation material. Sand undergoes some small immediate settlement and then stabilizes since it drains quickly. Clay generally is poor in bearing capacity. Clays do not drain quickly and will retain water for longer periods of time leading to long-term settlements. Most soils in reality are some combination of sands, clays, and silts which will behave somewhere in-between sand and clay. Exceeding the allowable bearing capacity of a soil will cause shear failure or excessive settlements. Bearing capacity is determined using the Terzaghi-Meyerhof equation:

$$q_{ult} = \frac{1}{2}\gamma B N_\gamma S_\gamma + cN_c S_c + (p_q + \gamma D_f)N_q$$

q_{ult} = Ultimate Bearing Capacity
γ = Soil Density
B = Width of Footing
c = Cohesion of Soil
N_γ = Density Bearing Capacity Factor
N_c = Cohesion Bearing Capacity Factor
N_q = Surcharge Bearing Capacity Factor
p_q = Surcharge Pressure
D_f = Depth from top of Soil to Bottom of Footing
S_γ = Density Shape Factor
S_c = Cohesion Shape Factor
The following table provides bearing capacity factors based on the internal angle of friction. In between values may be interpolated:

ϕ (Degrees)	N_c	N_q	N_γ
0	5.7	1.0	0
5	7.3	1.6	0.5
10	9.6	2.7	1.2
15	12.9	4.4	2.5
20	17.7	7.4	5.0
25	25.1	12.7	9.7
30	37.2	22.5	19.7
34	52.6	36.5	35.0
35	57.8	41.4	42.4
40	95.7	81.5	100.4
45	172.3	173.3	297.5
48	258.3	287.9	780.1
50	347.5	415.1	1153.2

Shape Factors are based on the geometry of the footing where B is the width and L is the length as below:

B/L	S_c
1.0	1.25
0.5	1.12
0.2	1.05
Strip Footing	1.00
Circular	1.20

B/L	S_γ
1.0	0.85
0.5	0.90
0.2	0.95
Strip Footing	1.00
Circular	0.70

The ultimate bearing capacity then needs to be corrected for overburden to find the net bearing capacity:

$q_{net} = q_{ult} - \gamma D_f$

The allowable bearing capacity is then determined by dividing the net capacity by a predetermined factor of safety. A factor of safety of between 2 and 3 is common:

$q_a = q_{net}/FS$

Slope Stability

There are 3 types of slope failures:

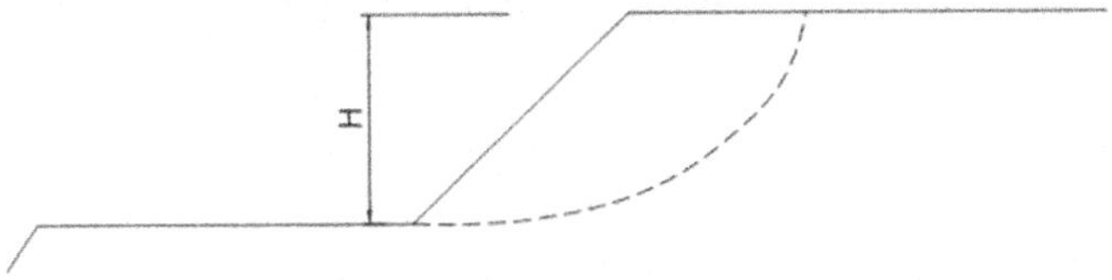
H

Toe Circle Failure

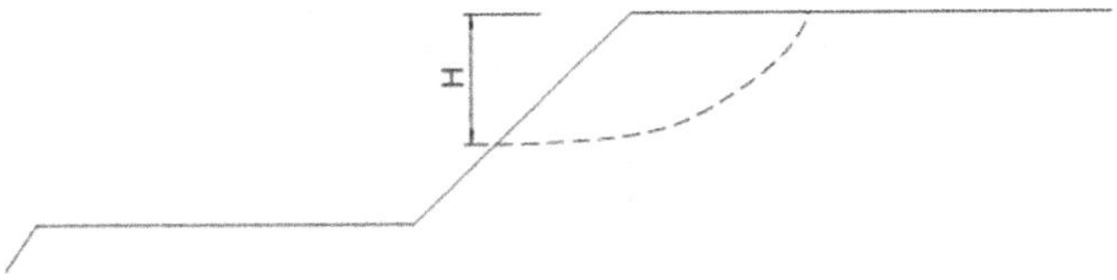
H

Slope Circle Failure

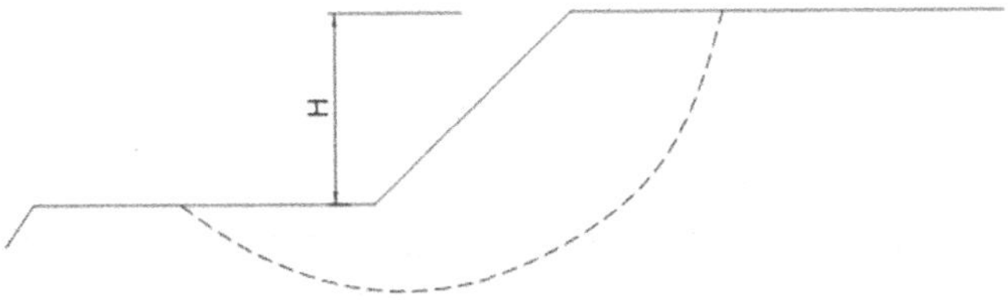
H

Base Circle Failure

$$F_{cohesive} = \frac{N_o c}{\gamma_{eff} H}$$

$F_{cohesive}$ = Safety factor for slope stability of cohesive soils. The minimum is often taken between 1.3-1.5

N_o = Stability number

c = cohesion (psi)

γ_{eff} = Effective soil density = $\gamma_{saturated}$ - γ_{water}

H = Depth of cut

Structural

Dead and Live Loads

Dead Loads: Loads which are permanent in the final condition of the structure. Examples include self-weight and additional permanent loads (such as pavement). Dead load factors are often lower than other types of loads. This is due to the higher level of reliability being able to predict the magnitude and character of these loads.

Live Loads: Loads which will or may change over time. In general, live loads represent pedestrian or vehicle loads. The load factor for live loads is often much higher due to the unpredictability.

In LRFD different types of loads are factored to represent a safety factor based on the reliability of our ability to accurately predict certain loading conditions. If only Dead and Live loads are present, the likely load combination is:

$1.2D + 1.6L$

Trusses

Trusses are structural members used to span long distances. Trusses are built up by members which are only in axial tension or axial compression. They can be analyzed by the method of joints as illustrated below. Consider the example truss with nodes labeled. To design, the axial force in each member must be determined. If we wanted to find the force in member BD, first like a typical beam, the reactions at A and B can be found by summing forces.

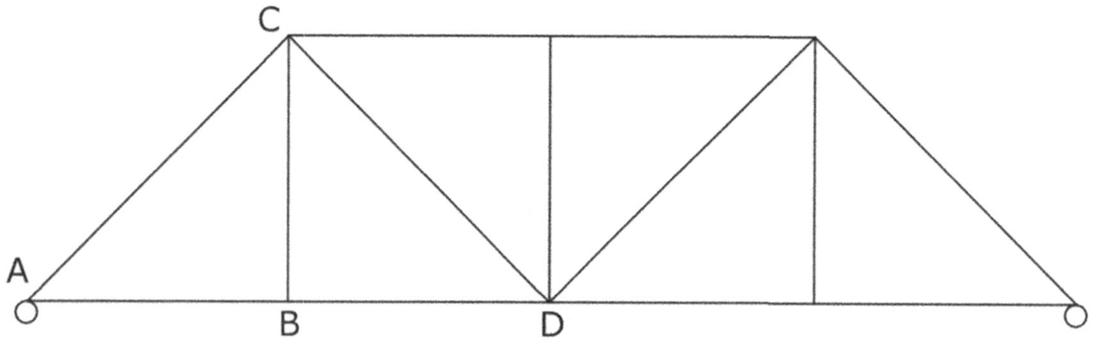
C
A
B
D

Then take a free body diagram of only the joint at A. This is shown below. In summing vertical reactions and since the reaction at A was found, the force in AC can be determined. Then there are only 2 horizontal forces and the force in AB can be found:

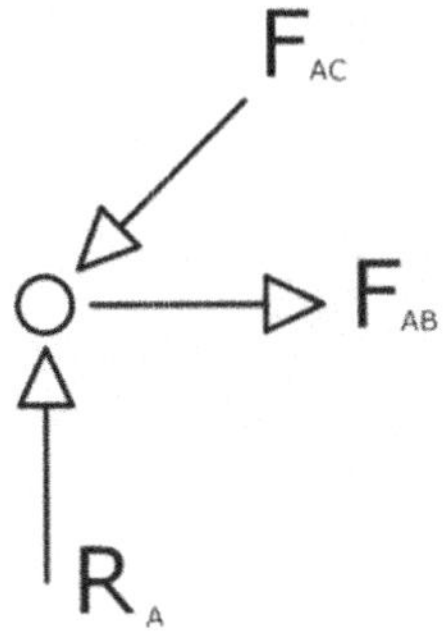

F AC
F AB
R A

Then take a free body diagram of Joint B as shown below. Since there are only two horizontal forces, the axial force in member BD can then be determined:

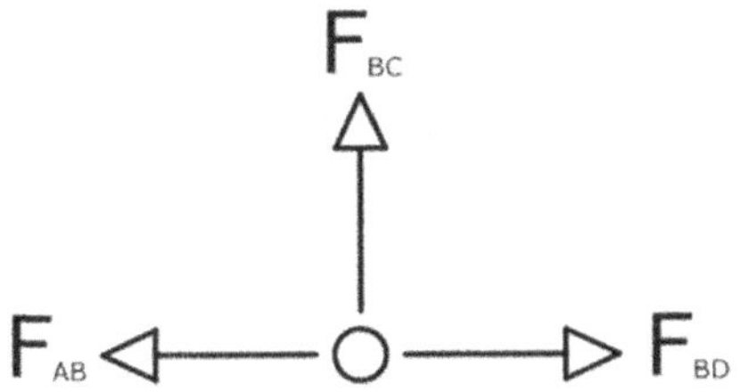

F BC
F AB
F BD

The remainder of the truss can be analyzed similarly.

Zero force members:

When determining how many 0-Force members a truss has, analyze each joint individually as a free body diagram and follow these guidelines:

1. In a joint with 2 members and no external forces or supports, both members are 0-force
2. In a joint with 2 members and external forces, If the force is parallel to one member and perpendicular to the other, then the member perpendicular to the force is a 0-force member.
3. In a joint with 3 members and no external forces, if 2 members are parallel then the other is a 0-force member

All other members are non-zero.

Bending Stress

Bending Stress

Mc/I

M = Applied Moment
c = Distance from the Centroid of the Cross Section to the Desired Location of Stress
I = Moment of Inertia of the Cross Section

Shear Stress

Shear stress at any point along a beam is the shear at that point over the area.

$\tau = V/A$

V = Shear at the point of interest
A = Cross sectional area

There is also horizontal shear stress due to bending

Horizontal shear stress $\tau = VQ/Ib$

V = Applied Shear Force (kips)
Q = First Moment of the Desired Area = ay.
a = Cross Sectional Area from Point of Desired Shear Stress to Extreme Fiber (in^2)
y = Distance from Centroid of Beam to Centroid of Area "a" (in)
I = Moment of Inertia of Beam (in^3)
b = Width of Member (in)

Axial Stress

Axial Stress:

P/A

P = Applied Force
A = Cross Sectional Area

Deflection

Deflection is the degree to which an element is displaced under load. Common equations for the maximum deflection of beams can be found in the Beam Chart

Beams

The chart below shows reactions, moments, and max deflections for common beam types:

Beam Type	Reaction	Maximum Moment	Maximum Deflection
Simply Supported w/ Single Point Load at Center	$\frac{P}{2}$	$\frac{PL}{4}$	$\frac{PL^3}{48EI}$
Simply Supported w/ Uniform Distributed Load	$\frac{wL}{2}$	$\frac{wL^2}{8}$	$\frac{5wL^4}{384EI}$
Cantilever w/ Single Point Load at Free End	P	PL	$\frac{PL^3}{3EI}$
Cantilever w/ Uniform Distributed Load	wL	$\frac{wL^2}{2}$	$\frac{wL^4}{8EI}$
One End Fixed, Supported at Other w/ Single Point Load at Center	$\frac{5P}{16}$, at Supported End $\frac{11P}{16}$, at Fixed End	$\frac{3PL}{16}$	$0.00932\frac{PL^3}{EI}$
One End Fixed, Supported at Other w/ Uniform Distributed Load	$\frac{3wL}{8}$, at Supported End $\frac{5wL}{8}$, at Fixed End	$\frac{wL^2}{8}$	$\frac{wL^4}{185EI}$
Beam Fixed at Both Ends, Single Point Load at Center	$\frac{P}{2}$	$\frac{PL}{8}$	$\frac{PL^3}{192EI}$
Beam Fixed at Both Ends, Uniform Distributed Load	$\frac{wL}{2}$	$\frac{wL^2}{12}$	$\frac{wL^4}{384EI}$

Shear and Moment Diagrams

Shear and moment diagrams are a graphical representation of the forces applied along the length of a beam. The following rules are used to develop shear diagrams:

- A concentrated force causes a jump in the shear diagram of equal magnitude
- A distributed load causes a line in the diagram with slope equal to the distributed load
- Forces up are positive and down is negative

This is depicted graphically below:

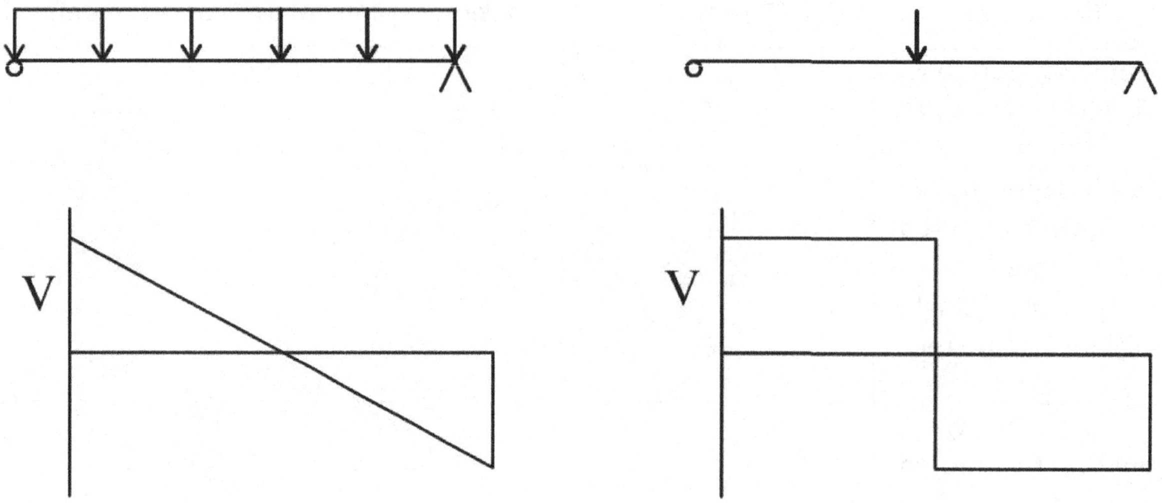
V
V

The following are rules for construction of a moment diagram:

- The moment at any point on the graph is equal to the area under the shear diagram up to this point
- An isolated moment causes a jump in the diagram of equal magnitude
- The shear at any point in the beam is equal to the slope of the same point on the moment diagram
- A distributed load will cause a parabolic moment diagram curve

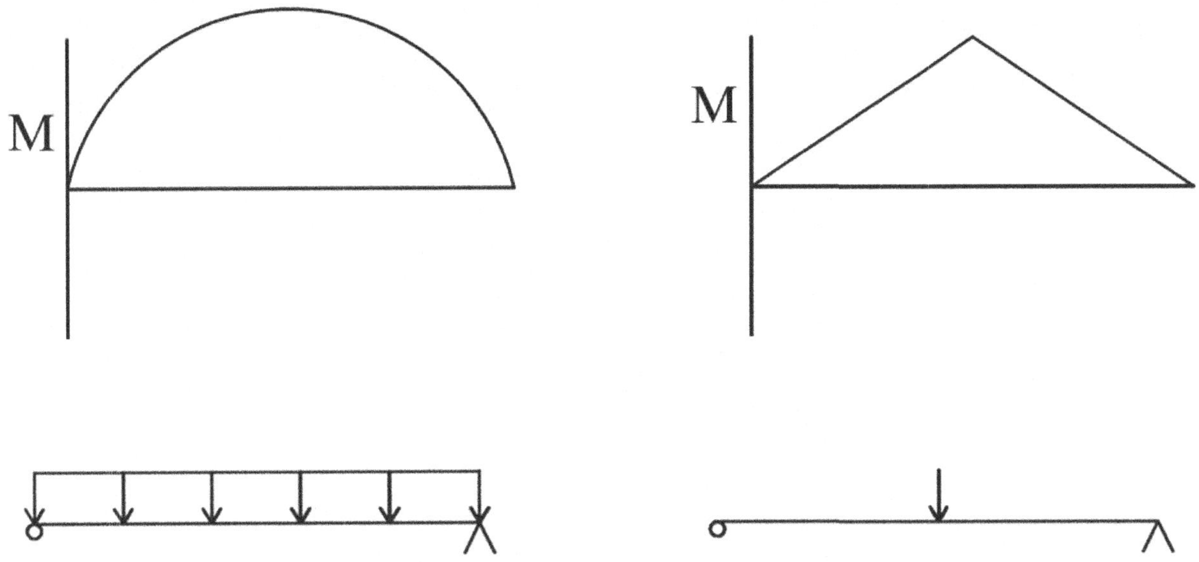
M
M

Columns

Columns are vertical members used to carry the load from spanning members to the foundation. For ideal columns where the load is concentric, the Euler formula is used to determine the theoretical maximum load:

Critical Load

$$P_{Cr} = \frac{\pi^2 EI}{(KL)^2}$$

E = Modulus of Elasticity (psi)
I = Moment of Inertia (in^4)
K = Effective Length Factor (See chart below)
L = Effective Length (in)

Critical Stress

$$F_{Cr} = \frac{\pi^2 E}{\left(\frac{KL}{r}\right)^2}$$

Effective Length Factor Chart

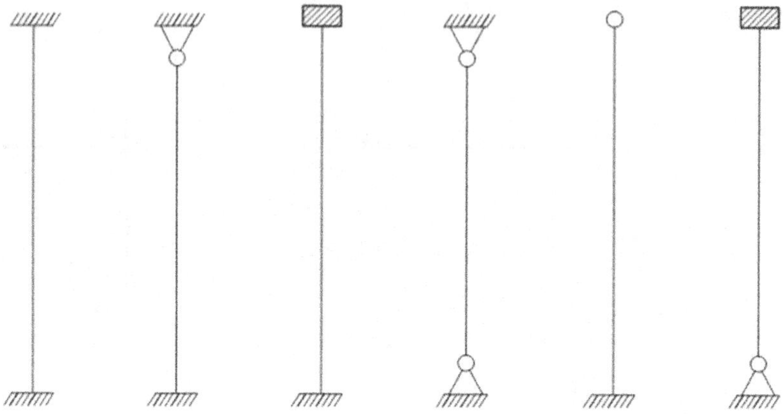

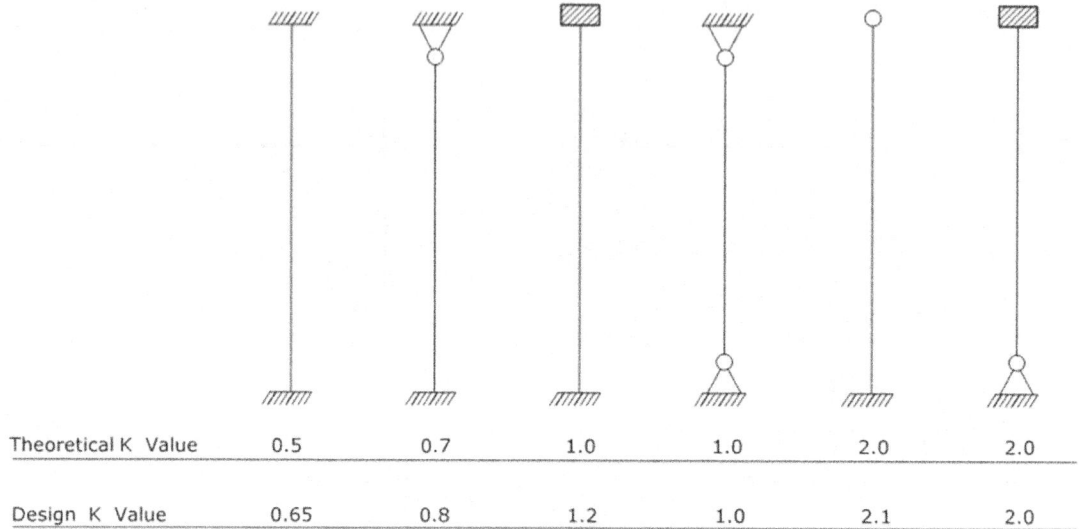

Theoretical K Value	0.5	0.7	1.0	1.0	2.0	2.0
Design K Value	0.65	0.8	1.2	1.0	2.1	2.0

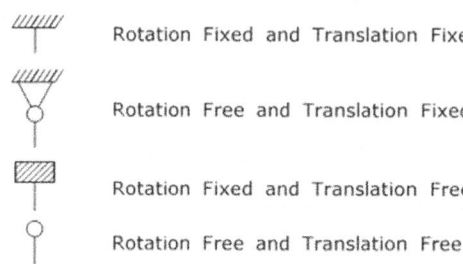

Rotation Fixed and Translation Fixed

Rotation Free and Translation Fixed

Rotation Fixed and Translation Free

Rotation Free and Translation Free

Slabs

One-way slabs:

Slabs are structural elements whose length and widths are large compared to the thickness. Slabs are often used as floors or as foundation elements.

Flexure:

Slabs must be analyzed by simplified methods due to the indeterminacy of a full analysis. The most common of which is to analyze as a 1-foot wide strip and treat the span length as a beam. Transverse reinforcement is necessary to control temperature and shrinkage.

Shear:

Shear in slabs is also determined by taking one-foot wide sections to analyze as a beam. However often shear will not control.

Footings

This section will focus on shear and moment for wall footings and column spread footings.

One-Way Shear:

The critical section for one-way shear is at a distance d from the support.

The force applied to the footing is assumed to spread uniformly therefore the bearing pressure q_u is the force divided by the area:

$$q_u = P/(Lb_w)$$

P = Factored Load
L = Length of Footing
b_w = Width of Footing

The shear force then is the shear resulting from the bearing pressure in the area from the critical section to the free end:

$$V_u = q_u b_w \left(\frac{L}{2} - \frac{a}{2} - d\right)$$

a = Width of Column
d = Depth to Flexural Reinforcing

Moment:

The critical section for moment is at the face of the column. The applied moment resulting from the bearing pressure is:

$$M_u = q_u \frac{\left(\frac{L}{2} - \frac{a}{2}\right)^2}{2} b_w$$

Retaining Walls

Retaining walls are built to facilitate an immediate change in elevation. Some uses are to support roadways or a need for a wide, level area to be formed from a sloping existing grade. Retaining walls are designed to resist lateral loads from active earth pressure (see geotechnical section for computation of these loads) and surcharge loads which is any additional load imposed on the soil above, which when close enough will cause an additional pressure on the load due to the distribution of this load through soil. The stem of retaining walls can be analyzed as a cantilever beam extending vertically from the footing. The footing is composed of the toe which is the portion on the side of the lower elevation of soil and the heel which is portion on the side of the higher elevation of soil. Retaining walls are analyzed on a per foot width:

$$Moment\ at\ base\ of\ stem = M_{stem} = R_{ah} y_a$$

R_{ah} = Horizontal Active Earth Pressure per ft Width
y_a = Eccentricity of Horizontal Active Earth Pressure

For shear, the critical section is a distance, d, from the base of the stem where d is the distance from the main flexural reinforcement (Heel side) to the extreme compression face (Toe side):

$$V_{stem} = R_{ah}$$

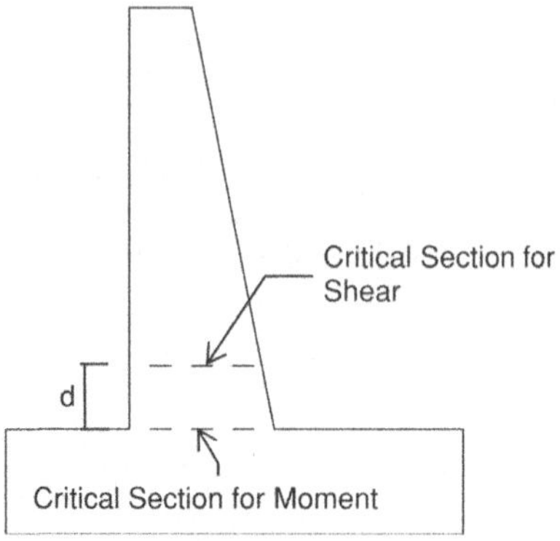
Critical Section for Shear
d
Critical Section for Moment

Hydraulics and Hydrology

Open Channel Flow

For open channel flow use the Chezy-Manning equation:

$Q = (1.49/n)AR^{2/3}S^{1/2}$

Q = Flow Rate (cfs)
n = Roughness Coefficient
A = Area of Water
R = Hydraulic Radius
S = Slope (decimal form)

The hydraulic radius is the area of water divided by the wetted perimeter which is the perimeter of the sides of the channel which are in contact with water.

Stormwater Collection

There are many components used in the collection of stormwater some examples include:

Culverts: A pipe carrying water under or through a feature. Culverts often carry brooks or creeks under roadways. Culverts must be designed for large intensity storm events.

Stormwater Inlets: Roadside storm drains which collect water from gutter flow or roadside swales.

Gutter/Street flow: Flow which travels along the length of the street or gutter.

Storm Sewer Pipes: Pipes installed under the road which carry the water from inlets to a suitable outlet.

The principle of Conservation of Flow is often applicable when analyzing drainage. It states that the flow in must equal a flow out and therefore:

$Q_1 + Q_2 = Q_3$

Storm Characteristics

Storm characteristics include duration, total volume, and intensity

Duration: The length of time of a storm. Often measured in days and hours.

Total Volume: The entire amount of precipitation throughout the duration of the storm in a defined area.

Storm Intensity: Total volume of the storm divided by the duration of the storm event. Intensities can be averaged over the entire storm or at shorter intervals to provide instantaneous high intensity portions of the storm. Hyetographs are bar graphs used to measure instantaneous rainfall intensities over time.

A design storm must be specified when performing any calculations. The design storm is defined by its recurrence interval which is the given amount of time it is likely to see a storm of a certain intensity. Design storms are often 10, 20, 50, or 100-year storms meaning a storm of a certain intensity would only occur once within the given duration.

Hyetographs – Graphical representation of rainfall distribution over time

Hydrograph – Graphical representation of rate of flow vs time past a given point often in a river, channel or conduit

Parts of a Hydrograph are shown graphically:

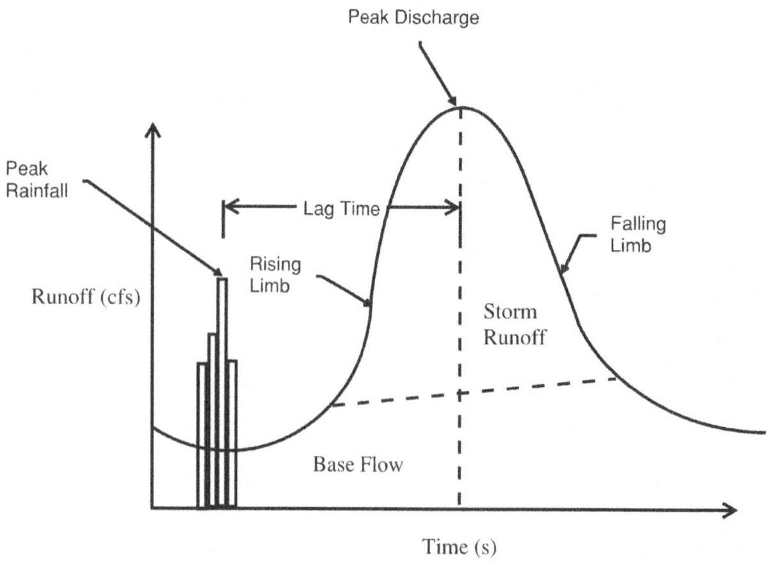

Peak Discharge
Peak Rainfall
Lag Time
Falling Limb
Rising Limb
Runoff (cfs)
Storm Runoff
Base Flow
Time (s)

Runoff Analysis

The rational method can be used to determine the flow rate from runoff of a drainage area. The equation is:

$Q = ACi$

Q = Flow Rate (cfs)
A = Drainage Area (Acres)
C = Runoff Coefficient
i = Rainfall Intensity (in/hr)

For total flow from multiple areas to a single outlet the conservation of flow principle is applied and the total is the sum of all flow into the outlet.

Time of Concentration, t_c: The time of travel for water to move from the hydraulically most remote point in a watershed to the outlet. The time of concentration is the sum of three components:

$$t_c = t_{sheet} + t_{shall} + t_{chann}$$

For approximately the first 300 ft, water moves as sheet flow:

$$t_{sheet} = \frac{0.007(nL_o)^{0.8}}{\sqrt{P_2}S_{deciaml}^{0.4}}$$

n = Manning's Roughness Coefficient
L_o = Length Overland
P_2 = 2-yr, 24-hour rainfall (in.)
$S_{decimal}$ = Slope of the Hydraulic Grade in Decimal Form

Flow then becomes shallow concentrated flow and typically enters a swale or ditch:

$$t_{shallow} = \frac{L_{shallow}}{v_{sha}}$$

$v_{shall} = 16.1345\sqrt{S_{decim}}$ for unpaved flow

$v_{shall} = 20.3282\sqrt{S_{decimal}}$ for paved flow

$L_{shallow}$ = Length of Concentrated Flow

$V_{shallow}$ = Velocity of Concentrated Flow

Flow finally will then enter a storm drain or channel:

$$t_{channel} = \frac{L_{chann}}{v_{chann}}$$

$L_{channel}$ = Length of Channel

$V_{channel}$ = Velocity of Channel (Often determined by Manning or Hazen-Williams Equation)

<u>NRCS/SCS Runoff Method</u>

This is an alternative method for determining runoff:

$$S = \frac{1000}{CN} - 10$$

S = Storage Capacity of Soil (in.)
CN = NRCS Curve Number

$$Q = \frac{\left(P_g - 0.2S\right)^2}{P_g + 0.8S}$$

Q = Runoff (in.)
P_g = Gross Rain Fall (in.)

<u>Detention/Retention Ponds</u>

Detention and retention ponds are often used to collect water for flood control and stormwater runoff treatment.

Detention Ponds: Also known as dry ponds. These are ponds which are often kept dry except during flood events. The pond will fill up during increased precipitation to control the flow intensity. This is common in dry, arid or urban areas to prevent excessive flooding. The ponds typically will be designed to hold water for about 24 hours. Detention ponds also controls the amount of sediment since it is captured in the pond and then typically becomes accessible after the pond has dried.

Retention Ponds: Also called wet ponds since they contain a volume of water at all times. The elevation of the water will vary depending on precipitation but will always maintain a permanent amount of water based on low flow conditions. This allows sediment control since the deposits will settle to the bottom and allow for collection.

Pressure Conduits

The Darcy Equation is used for fully turbulent flow to find the head loss due to friction. The equation is:

$h_f = (fLv^2)/(2Dg)$

h_f = Head Loss Due to Friction (ft)
f = Darcy Friction Factor
L = Length of Pipe (ft)
v = Velocity of Flow (ft/sec)
D = Diameter of Pipe (ft)
g = Acceleration Due to Gravity, Use 32.2 ft/sec^2

The Hazen-Williams equation is also used to determine head loss due to friction. Be aware of units as this equation may be presented in different forms. The most common is the following:

$h_f = 10.44LV^{1.85}/C^{1.85}d^{4.87}$

h_f = Head Loss due to Friction (ft)
L = Length (ft)
V = Velocity (gallons per minute)
C = Roughness Coefficient
d = Diameter (in)

In addition to these losses, there is also what is called minor losses of energy due to friction

Minor Losses – Friction losses due to fittings in the line, changes in the dimensions of the pipe, or changes in direction

- Minor losses can be calculated as per the Method of Loss coefficients.
- Each change in the flow of pipe is assigned a loss coefficient, K
- Loss coefficients for fittings are most often determined and provided by the manufacturer

Loss coefficients for sudden changes in area must be determined:

For Sudden Expansions:

$$K = \left(1 - \left(\frac{D_1}{D_2}\right)^2\right)^2$$

For Sudden Contractions:

$$K = 0.5\left(1 - \left(\frac{D_1}{D_2}\right)^2\right)$$

$$D_1 = Smaller\ diamter\ pipe$$

Loss coefficients are then multiplied by the kinetic energy to determine the loss.

$$h_f = K\frac{v^2}{2g}$$

Bernoulli Equation

The Bernoulli equation for the conservation of energy states that the total energy is equal to the sum of the pressure + kinetic energy + potential energy and is conserved at any point in the system. Therefore:

$E_t = E_{pr} + E_v + E_p = p + v^2/2g + z$

E_{pr} = Pressure = p
E_v = Kinetic Energy = $v^2/2g$
v = Velocity (ft/s)
g = Acceleration Due to Gravity (32.2 ft/s^2)
E_p = Potential Energy = z = Height above point of interest to surface of water (ft)

Transportation and Geometrics

Horizontal curves

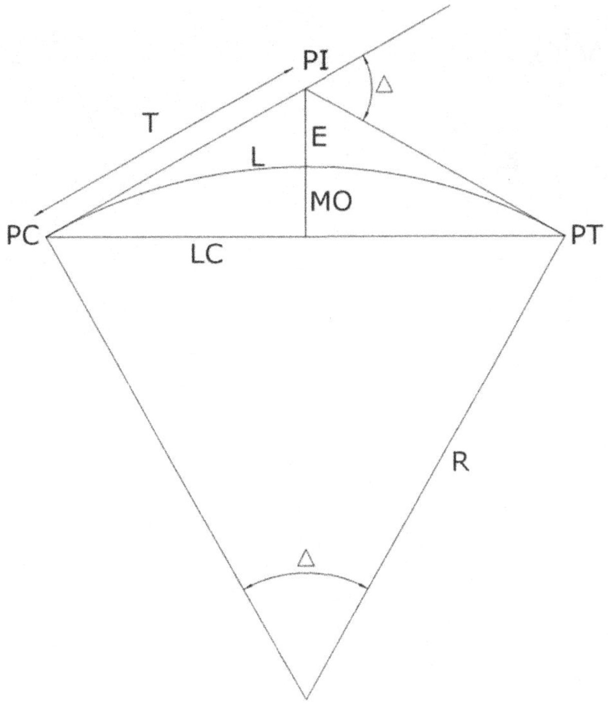

PI
Δ
T
E
L
MO
PC
PT
LC
R
Δ

$$\Delta = \frac{180}{\pi} \times \frac{L}{R}$$

$$L = \frac{\Delta \times \pi \times R}{180}$$

$$R = \frac{180 \times L}{\Delta \times \pi}$$

$$T = R \times \left(\tan\frac{\Delta}{2}\right)$$

$$E = T \times \left(\tan\frac{\Delta}{4}\right)$$

$$LC = 2 \times R \times \left(\sin\frac{\Delta}{2}\right)$$

$$MO = R \times \left(1 - \cos\frac{\Delta}{2}\right)$$

Vertical Curves

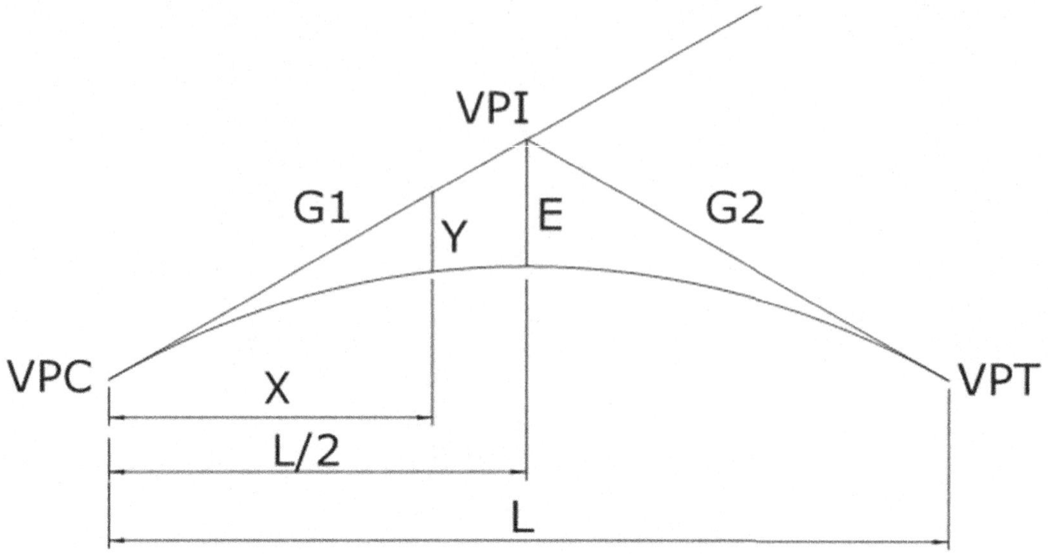
VPI
G1
Y
E
G2
VPC
VPT
X
L/2
L

A = Gradient of the curve = $g_2 - g_1$
E = AL/800
Y = $Ax^2/200L$
Elevation at a distance x from VPC = $E_{VPC} + g_1x + ½(rx^2)$

Traffic Volume

The following are traffic volume factors. It is important to note for how many lanes and directions the values represent:

Average Daily Traffic, ADT: Average number of vehicles per day over a given time period.

Average Annual Daily Traffic, AADT: Average number of vehicles per day over a year. Typically, it is calculated by dividing the total volume of vehicles in a year by 365 days.

Average Daily Truck Traffic, ADTT: Average number a trucks per day in a given time period.

Design Hour Volume, DHV: The hour of volume used in design.

K-factor, K: The ratio of the Design Hour Volume to the Average Annual Daily Traffic (DHV/AADT).

Directional Factor, D: Percentage of volume for the dominant direction of traffic during peak flow.

Directional Design Hour Volume, DDHV: The product of the Directional Factor and the Design Hour Volume.

Rate of Flow, v: Equivalent hourly rate at which vehicles pass a given point during a given time interval. The time frame is often taken as 15 min.

Design Capacity: Maximum volume a given roadway can handle.

Ideal Capacity, c: Ideal amount of volume for a given roadway. For freeways this is often taken as 2400 passenger cars per hour per lane (pcphpl).

Volume to capacity ratio: Volume over capacity, v/c

Peak Hour Factor, PHF: Ratio of the peak hour volume to the peak rate of flow in that hour:

$$PHF = \frac{V_{vph}}{4V_{15\ min,peak}}$$

Vehicle Dynamics

The distance it takes a driver to stop after recognizing an obstruction is the sum of two components. The first is before breaking and the second is after breaking.

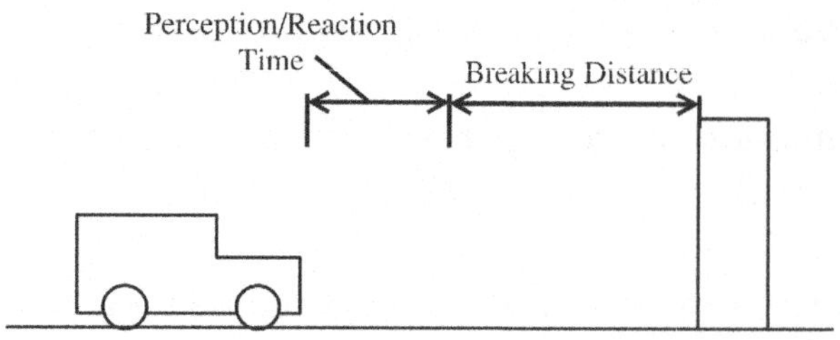

Perception/Reaction Time
Breaking Distance

This is represented in the following equation:

$S_{stopping} = vt_p + s_b$

The first component assumes the velocity is constant during the perception reaction time which is the time it takes the driver to recognize the obstruction and begin the breaking process. This is calculated by:

vt_p where

v = Velocity (ft/s)
t_p = Perception Reaction Time (seconds)

The stopping distance during the breaking process is the following:

$s_b = v^2_{mph}/(30(f+G))$

v = Velocity (miles per hour)
f = Friction Factor with the Pavement
G = slope (positive for uphill and negative for downhill)

To convert the velocity from ft/s to mi/hour where applicable multiply by:

(3600 s/hr)/(5280 ft/mi)

Materials

Soil Classification

There are two types of common soil classification system. First is the AASHTO:

	Granular Materials (35% or less passing no. 200 sieve)							Silt-Clay materials (more than 35% passing no. 200 sieve				
	A-1		A-3	A-2				A-4	A-5	A-6	A-7	A-8
	A-1-a	A-1-b		A-2-4	A-2-5	A-2-6	A-2-7				A-7-5 or A-7-6	
Sieve analysis: % passing:												
no. 10	50 max											
no. 40	30 max	50 max	51 min									
no. 200	15 max	25 max	10 max	35 max	35 max	35 max	35 max	36 min	36 min	36 min	36 min	
Characteristics of fraction passing no 40:												
LL: liquid Limit				40 max	41 min	40 max	41 min	40 max	41 min	40 max	41 min	
PI: Plasticity Index	6 max		NP	10 max	10 max	11 min	11 min	10 max	10 max	11 min	11 min	
Usual types of significant constituents	Stone fragments gravel and sand		Fine Sand	Silty or clayey gravel and sand				Silty Soils		Clayey Soils		Peat, highly organic soils
General subgrade rating	Excellent to good					Fair to poor						Unsatisfactory

Second is the Unified Soil Classification System (USCS):

Major Division		Group Symbol	Laboratory classification criteria		Soil Description
			% finer than 200 sieve	Supplementary requirements	
Coarse-grained (over 50% by weight coarser than no. 200 sieve)	Gravelly soils (over half of coarse fraction larger than no. 4)	GW	0-5[a]	D_{60}/D_{10} greater than 4. D^2_{30} % / $(D_{60}D_{10})$ between 1 and 3	Well graded gravels, sandy gravels
		GP	0-5[a]	Not meeting above gradation requirement for GW	Gap-graded or uniform gravels, sandy gravels
		GM	12 or more[a]	PI less than 4 or below A-Line	Silty gravels, silty-sandy gravels
		GC	12 or more[a]	PI over 7 and above A-Line	Clayey gravels, clayey sandy gravels
	Sandy soils (over half of coarse fraction finer than no. 4)	SW	0-5[a]	D_{60}/D_{10} greater than 6. D^2_{30} % / $(D_{60}D_{10})$ between 1 and 3	Well-graded, gravelly sands
		SP	0-5[a]	Not meeting above gradation requirement for SW	Gap-graded or uniform sands, gravelly sands
		SM	12 or more[a]	PI less than 4 or below A-Line	Silty sand
		SC	12 or more[a]	PI over 7 and above A-Line	Clayey sands, clayey gravelly sands
Fine-grained (over 50% by weight finer than no. 200 sieve)	Low compressibility (liquid limit less than 50)	ML	Plasticity chart		Silts, very fine sands, silty or clayey fine sands, micaceous silts
		CL	Plasticity chart		Low plasticity clays, sandy or silty clays
		OL	Plasticity chart, organic odor or color		Organic silts and clays of high plasticity
	High Compressibility (liquid limit 50 or more)	MH	Plasticity chart		Micaccous silts, diatomaceous silts, volcanic ash
		CH	Plasticity chart		Highly plastic clays and snady clays
		OH	Plasticity chart, organic odor or color		Organic silts and clays of high plasticity
Soils with fibrous organic matter		Pt	Fibrous organic matter; will char, burn, or glow		Peat, sandy peats, and clayey peat

[a]For soils having 5-12% passing the no.200 sieve, use a dual symbol such as GW-GC

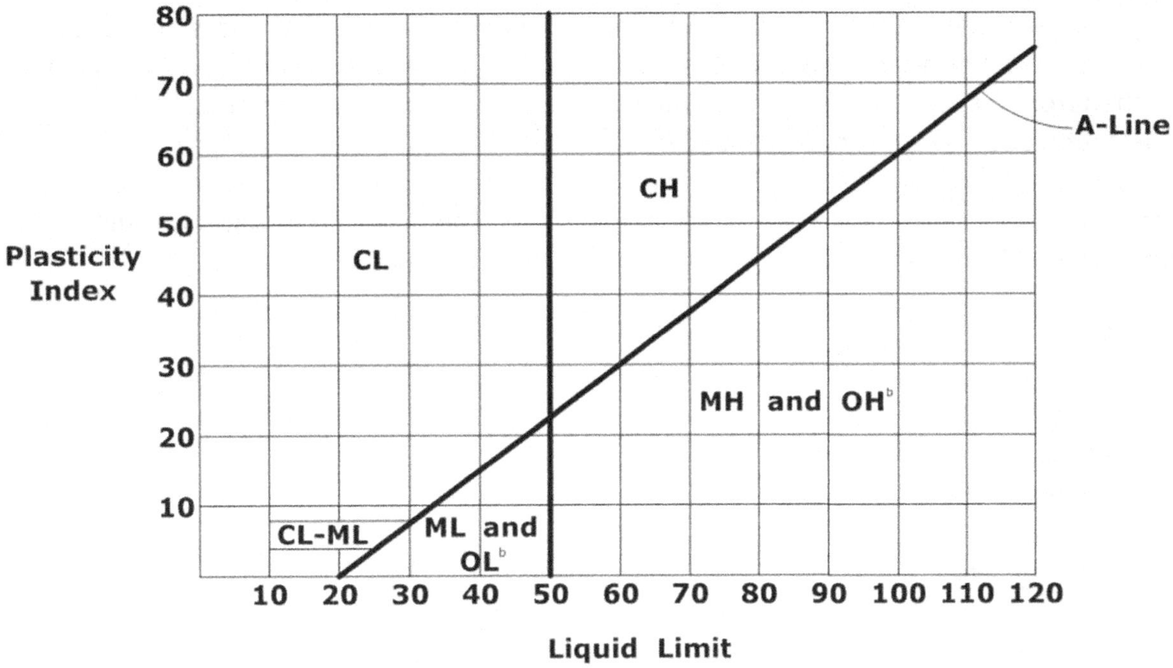
Plasticity Index
80
70
60
50
40
30
20
10
CL
CH
A-Line
MH and OH[b]
CL-ML
ML and OL[b]
10 20 30 40 50 60 70 80 90 100 110 120
Liquid Limit

Soil Properties

The strength of soil is often determined by the standard penetration test. This measures the resistance to penetration using a standard split spoon sampler which is hit by a 140 lb hammer dropped from 30" high. The number of blows required to drive the sampler 12" after an initial penetration of 6" is referred to as the N-Value.

Permeability of a soil is a measure of continuous voids. The flow rate of water through soil depending on its permeability can be measured by Darcy's Law:

$Q = KiA_{gross}$

Q = Flow rate (cfs)
K = Coefficient of permeability (ft/sec)
i = Hydraulic Gradient

Phase Relationships

W_V = Weight of voids
W_w = Weight of water
W_s = Weight of solids
W_T = Total weight
V_V = Volume of voids
V_w = Volume of water
V_s = Volume of solids
V_T = Total Volume

Void ratio = $e = V_V/V_s$
Porosity = $n = V_V/V_T$
Degree of Saturation = $S = V_w/V_v \times 100\%$
Moisture content = $w = W_w/W_s$
Dry Unit weight = density = γ = Weight/Volume
SG = Specific Gravity = γ_s/γ_w
$\gamma_w = 62.4\ lb/ft^3$ (constant)

A saturated sample indicates the volume of the voids = volume of water
A dry sample indicates the volume of voids includes no water

Concrete Properties

Concrete consists of cement, coarse aggregate, fine aggregate, and water. Additionally, concrete may contain admixtures to enhance a certain desirable aspect of the target product. Some properties of concrete include:

Concrete Strength, f'_c: The design compressive strength of the concrete. In general, this will range from 3000 to 6000 psi. However, strengths can be much higher such as 20,000 psi with proper mixing and additives.

Modulus of Elasticity (Normal weight concrete), $E_c = 57{,}000\sqrt{f'_c}$

Modulus of Rupture or the tensile strength, $f_r = 7.5\sqrt{f'_c}$ While the tensile strength of concrete is ignored in flexure, this is often used in cracking analysis.

Water to Cement ratio, w/c, is the amount of water to the amount of cement in a given mix. In general, the w/c ratio is inversely proportional to the strength since the higher amount of cement, the stronger the mix.

Cement Types:

Type I – General use cement. When special properties are not desired Type I can be used.

Type II – Used in areas where sulfate attack is a concern. This is often in areas exposed to groundwater such as drainage structures. Type II will cure at a slower rate and therefore produce less heat than other types and gain strength at a slower rate.

Type III – High early strength concrete. As opposed to type II or IV, a large amount of heat is released quickly and therefore is not suitable for mass-type pours. Type III is used in concrete where rapid strength gain is desirable such as precast concrete.

Type IV – Low heat of hydration. Gains strength slowly and generates a low amount of heat. Often used for mass-pours such as mat foundations or large retaining walls.

Structural Steel Properties

Yield Strength, F_y: Stress at which the steel will yield and begin to cause permanent deformations.

Ultimate Strength, F_u: Stress at which the steel will fracture or fail in brittle behavior.

Modulus of Elasticity, E_s = The tendency of a material to deform when subjected to forces. Also, is the ratio of stress over strain. Often in structural steel it is assumed to be taken as 29,000 ksi

Ductility: Measure of a materials ability to deform before failure. Ductility is the ratio of ultimate failure strain to yielding strain.

Toughness: The ability to withstand high stresses without fracturing.

Hardness: The ability of a material to resist surface deformation.

Material Test Methods and Specification Conformance

Concrete:

Strength tests: Most often strength is determined by loading cylinders often 6" in diameter to failure and recording the results.

Slump Test: Measure of the consistency and workability of a batch of concrete. A cone about 6" in diameter on the wide end and 12" tall is filled with concrete. The filled cone is placed on the ground and then removed to allow the concrete to naturally disperse. The remaining height and diameter of the concrete mix is measured and recorded.

Steel:

Tensile Test: Axially loading a steel member to record the strain in the member as the load increases. From this test the yield strength, ultimate strength, and stress strain curve can be determined. When the applied stress exceeds the yield strength, the member will undergo plastic deformation and the cross-sectional area will reduce until the member fractures. This is known as necking.

Fatigue Testing: Fatigue is damage caused by repeated cycles of loading. Even though the stress in fatigue is less than the yield strength of the member, the repetition over a long period of time can cause failure. A fatigue test measures the ability of a member to resist repeated cycles of stress at a given magnitude.

Scratch Hardness Test: Also known as Mohs Test. Compares the hardness of a material to that of minerals. Minerals of known and increasing hardness are used to scratch the sample and results are observed.

Charpy V-Notch Test: Measure of a member's toughness. A member is given a 45 degree notch and a pendulum is used to hit the opposite side of the member. This is performed at different heights and magnitudes until the member fails.

Compaction

Compaction is the reduction of voids in a mass of soil. The more compacted a mass of soil is, the more stable and strong it is to support a structure. Compaction is done by placing soil in layers called lifts and using equipment to mechanically apply weight and potentially vibration to the lifts. Some types of compaction equipment are Grid Rollers for rocky soil, sheep foot rollers for cohesive soils, or roller compactors with vibration capabilities for cohesion less soils.

When soil is compacted, the volume decreases. This is referred to as shrinkage. To calculate the compacted volume of a soil mass from its volume in its natural state use the following equation:

$$V_c = \left(\frac{100\% - \%shrinkage}{100}\right) V_b$$

V_c = Compacted Volume
%shrinkage = Percent Shrinkage
V_b = Volume of Soil in its Natural State

Conversely, when soil is removed from the ground there is an expansion of volume known as swelling:

$$V_l = \left(\frac{100\% + \%swell}{100}\right) V_C$$

V_l = Loose Volume
%swell = Percent Swell

Site Development

Excavation (Cut/Fill Estimates)

The most common method for determining the volume of excavation for cut and fill is the average end area method:

$V = L(A_1+A_2)/2$

L = Distance Between Area 1 and 2 (ft)
A_1 and A_2 = Respective Cross-Sectional Areas (ft^2)

Construction Site Layout

Construction sites are surveyed and markers are placed to indicate measurements and control points. These points are designated in the field by the use of stakes. These stakes can be called construction stakes, alignment stakes, offset stakes, grade stakes, or slope stakes depending on what they are meant to indicate. The accuracy of dimensions depends on the intent. Some accuracy requirements are shown below:

Type of Measurement	Level of Accuracy (ft)
Roadway Alignment, Intersections, and Paving	0.01
Buildings and Bridges	0.01
Culvert Lengths	0.1
Grade Staking	0.1
Culvert Stations	1.0
Telephone or Power Poles	1.0

Soil Erosion and Sediment Control

During construction activities involving excavation, there can be a significant amount of soil erosion leading to a dispersion of sediment. This needs to be controlled to prevent a negative impact to the surrounding areas. There are a number of options for sediment control:

Silt Fences: Fences consisting of a geotextile fabric and posts which allow the passing of runoff water but will catch the suspended sediment. They will be placed at the bottom of slopes and/or at the perimeter of the job sites at low points.

Hay Bales: Placed at the toe of slopes to help control runoff. Bales should be embedded in the ground and anchored securely with wooden posts.

Erosion Control Fabric: Geotextile fabric used for the control of erosion on steep slopes. Often these are used on piles of excavated material.

Temporary Seeding and Mulch: This involves seeding and mulching slopes to create growth that can control erosion due to the roots holding together soil. This is often used as a permanent measure for cut slopes.

Slope drain: A drain constructed to direct water to a specified area. The drain can be constructed with numerous materials such as plastic or metal pipes and concrete or asphalt. Drains must be properly anchored to resist forces from the flow of water. The outlet often is required to slow the flow of water by using energy dissipaters such as riprap.

Sediment Structure: An energy dissipating structure often made of rocks used to slow the flow of water and catch sediment.

Temporary Berm: A hill constructed of compacted soil to prevent runoff flowing in a specific direction. Berms are placed either at the top or bottom of slopes.

Impact of Construction on Adjacent Facilities

Construction can have a negative effect on surrounding properties and areas. These issues need to be anticipated and mitigated as possible. Some of the concerns include:

Construction Noise: OSHA sets maximum decibel limits on daily sound exposure. In the United States, this is typically 90 dBA for the eight-hour noise level

Runoff and Sediment: Construction sites, especially those involving excavation, can change the dynamics of runoff and drainage. See the section on Soil Erosion and Sediment Control for more details.

View: Construction projects often change the landscape of the affected area. This may have an impact on the look and feel of an area. The needs of adjacent properties may need to be considered for these changes.

Rights of Way: Often, land which is not owned by the owner of the project is necessary for the final or temporary conditions. In these cases, land needs to be acquired temporarily or permanently to complete the work. The owner of the project and of the land must come to an agreement to allow use of the property

Economic or social impact: Construction during and after may impact the access or desirability of a business or residential area. Consideration should be taken to limit the impacts to businesses or residents. For example, a bridge detour may cut off access to a restaurant which collects patrons mostly from tourists passing the effected route. The owner would then be compensated for the loss of business.

Safety

Safety is extremely important for construction sites. The *OSHA CFR 29 Part 1910 and Part 1926: Occupational Safety and Health Administration* provides requirements for all types of construction situations and is recommended to use for the exam. Some of the highlights which you should further familiarize yourself with include:

- Excavation Safety: Except for excavations in rock, anything deeper than 5 ft must be stabilized to prevent cave-in. This may be achieved by providing appropriate earth retention systems or sloping at appropriate rates. This is determined by the depth of excavation, soil type, and other requirements.
- Fall protection: Drop-offs must be protected from fall based on the height of the drop. Some examples of protection include temporary fences, nets, or lifelines.
- Roadside Safety: Construction sites adjacent to traffic must be sufficiently protected from impact. At higher speeds concrete barriers may be needed also known as temporary precast concrete barrier curbs (TPCBC). At lower speeds it may be acceptable to provide barrels or cones to delineate the work area.
- Power line Hazards: For power lines which are electrified, all construction activities must be a minimum distance from the lines. This is based on the voltage of the lines. Typically the safe operational distance is 10 ft. for lines less than 50 kV and typically 35 ft. for lines greater than 50 kV.
- Confined Spaces: Anyone required to enter confined spaces must be appropriately trained and equipped. Oxygen must be monitored and kept at an acceptable level.
- Personal Protective Equipment (PPE): Equipment required by any personnel present on a job site. The main aspects are acceptable head protection and steel toed shoes.

STRUCTURAL DEPTH

American Society of Civil Engineers (ASCE-7)

Live Loads

Design live loads are based on the type of building (Occupancy) and can be determined from Table 4-1

In applicable cases for columns, live loads can be reduced by the following equation:

$L = L_O(0.25 + 15/\sqrt{K_{LL}A_T})$ if $A_T K_{LL} > 400$
L_O = Live load per occupancy
K_{LL} = Live load element factor, see Table 4-2
A_T = Tributary Area

However, the reduced live load must be greater than $0.4L_O$.

Live loads may be reduced if the following requirements are met:

- Tributary area must be greater than 400 ft^2
- Live Loads > 100 psf shall not be reduced (Exception: Live loads for members supporting two or more floors may be reduced by 20 percent)
- The live loads for passenger car garages shall not be reduced (Exception: Live loads for members supporting two or more floors may be reduced by 20 percent)
- Live loads of 100 psf or less in public assembly occupancies shall not be reduced

Snow Loads

Snow loads are determined from Chapter 7. There are 3 types of snow loads to know: ground, flat roof, and design roof.

P_g = ground snow load. This is the snow load on the ground as per a specific geographic area.

P_f = flat roof snow load = $P_g(0.7C_eC_tI_S)$

C_e = Exposure Factor from Table 7-2
C_t = Thermal Factor from Table 7-3
I_S = Importance factor from 1-5.2

Ensure this is greater than the minimum = $20I_S$

The design roof snow load $P_S = C_SP_f$

Determine C_S from Table 7-2c .

Snow Drift

Snow drift is additional load due to snow building up against a vertical wall from wind. The additional load is approximated by a triangular cross section of snow. Figure 7-8 depicts the necessary variables. To solve:

First determine the density of snow:

$\gamma = 0.13p_g + 14 < 30$ pcf

Then using the density, you can determine the height of the roof snow $h_b = p_s/\gamma$

h_c = the vertical distance from the top of roof snow to upper roof = height to upper roof - h_b
If $h_c < 0.2h_b$, Snow drift does not need to be applied

Then using figure 7-9 determine the drift height h_d which will be the larger of:
For Leeward drift use l_u = the length of the upper roof
For Windward use l_u = the length of the lower roof and only use ¾ of the h_d as determined from figure 7-9

Then calculate the width of the drift, w, for $h_d < h_c$ w=4h, if $h_d > h_c$ $w = 4h_d^2/h_c$ however w shall not be greater $8h_c$

The variables are better depicted in the diagram below:

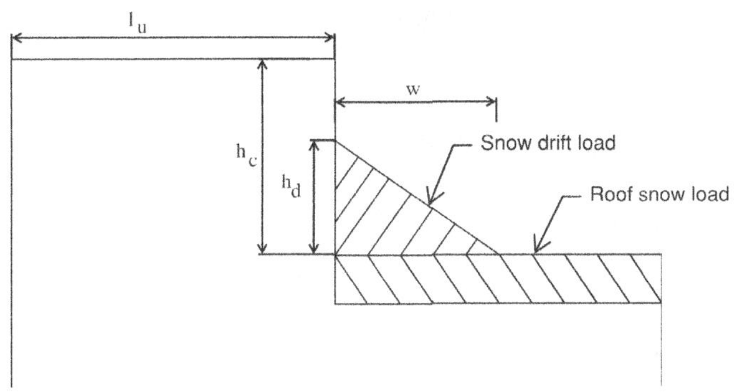
l_u
w
h_c
h_d
Snow drift load
Roof snow load

Site Classification and Occupancy

- From Table 20.3-1 the site classification can be determined
- Table 1.5-2 for the risk category

Site Coefficients and Spectural Response Modification Factors

S_{DS} = Design Spectural response acceleration parameter at short periods = $2/3S_{MS}$ = $2/3F_aS_S$
S_{MS} = The risk targeted maximum considered earthquake ground motion acceleration parameter (MCE_R) Spectural response acceleration parameter at short periods.
F_a = Site coefficient defined in table 11.4-1
S_S = Mapped (MCE_R) Spectural response acceleration parameter at short periods determined in accordance with section 11.4.1

S_{D1} = Design Spectural response acceleration parameter at a period of 1 s = $2/3S_{M1}$ = $2/3F_vS_1$
S_{M1} = The risk targeted maximum considered earthquake ground motion acceleration parameter (MCE_R) Spectural response acceleration parameter at a period of 1 s.
F_v = Site coefficient defined in table 11.4-1
S_1 = Mapped (MCE_R) Spectural response acceleration parameter at a period of 1 sec as determined in accordance with section 11.4.1

From Table 20.3-1 the site classification can be determined

Determine the Seismic design category based on short period response acceleration parameters from Table 11.6-1

Determine the Seismic design category based on 1-S period response acceleration parameters from Table 11.6-2

Effective Seismic Weight

Effective weight is the load which can be accounted for to offset horizontal seismic forces. This includes the dead load and any additional loading as outlined in section 12.7.2 such as:

1. In areas used for storage, a minimum of 25 percent of the floor live load (floor live load in public garages and open parking structures need not be included).
2. Where provision for partitions is required by Section 4.2.2 in the floor load design, the actual partition weight or a minimum weight of 10 psf (0.48 kN/m2) of floor area, whichever is greater.
3. Total operating weight of permanent equipment.
4. Where the flat roof snow load, P_f, exceeds 30 psf (1.44 kN/m2), 20 percent of the uniform design snow load, regardless of actual roof slope.

Seismic Base Shear and Force Distribution

Equivalent Lateral Force Procedure Section 12.8

The seismic base shear by the equivalent force method $V = C_sW$

C_S = Seismic response coefficient = $S_{DS}/(R/I_e)$

Determine I_e from table 1.5-2 using the risk category

R = response modification factor from table 12.2-1

The lateral seismic force at a given level shall be:

$$F_x = C_{vx}V$$

$$C_{vx} = \frac{w_x h_x^k}{\sum_{i=1}^{n} w_i h_i^k}$$

C_{vx} = vertical distribution factor

V = total design lateral force or shear at the base of the structure (kip or kN)

w_i and w_x = the portion of the total effective seismic weight of the structure (W) located or assigned to Level i or x

h_i and h_x = the height (ft) from the base to Level i or x

k = an exponent related to the structure period as follows:
for structures having a period of 0.5 s or less,
k = 1 for structures having a period of 2.5 s or more,
k = 2for structures having a period between 0.5 and
2.5 s, k shall be 2 or shall be determined by linear
interpolation between 1 and 2

AASHTO

Introduction to Bridges

Parts of a bridge include:

- Foundation – Structural members used to transfer load to the supporting soil.
- Substructure – Structural parts that support the horizontal span
- Superstructure – Structural parts which provide the horizontal span

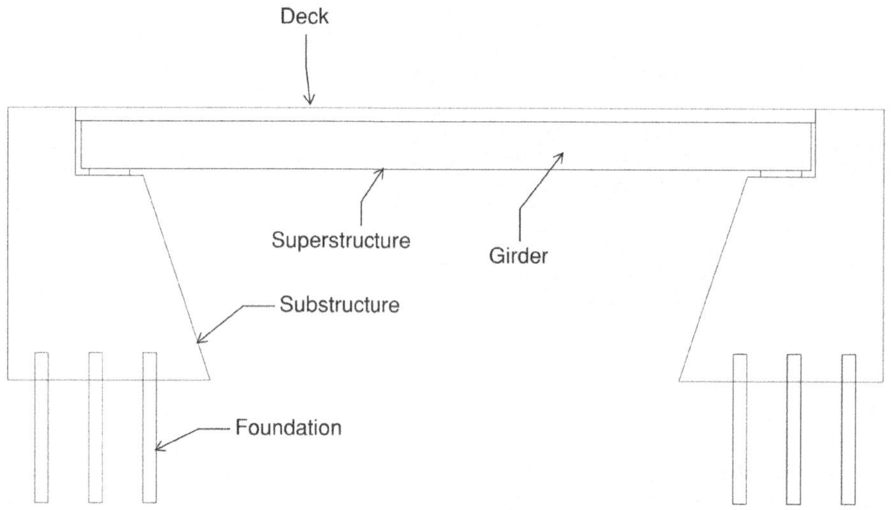
Deck
Superstructure
Girder
Substructure
Foundation

Limit States and Load Factors

Load factors and load combinations are handled differently in AASHTO. Different loading conditions are represented by Limit States. Some examples are Strength I, Strength III, and Service I. The load factors vary in magnitude depending on which limit state is applied. The load factors are then multiplied by the various types of loads. The Load Factors are found in Tables 3.4.1-1 and 3.4.1-2.

Live Load Distribution

Live load on bridges is not distributed evenly to girders. Live load distribution provides a more appropriate distribution based on girder spacing, deck thickness, type of bridge etc. The applicable cross sections are from Table 4.6.2.2.1-1

Then the appropriate equation on pages 4-37 through 4-45 determines the distribution of load. Be aware of the appropriate mode of failure and whether the beam is interior or exterior

American Concrete Institute (ACI)

Flexure

Moment capacity in concrete beams is based on the tension in the member being equal to the compression. The moment capacity then is the area of steel multiplied by the strength of steel multiplied by the distance from the steel centroid to the centroid of the compression block. Therefore:

$\phi M_n = \phi A_S F_Y(d-a/2)$, $\phi = 0.9$ for tension-controlled sections

A_s = area of steel (in^2)
F_y = yield strength of steel (ksi)
d = depth of tension steel (in)
a = depth of compression block (in)

And since Tension = Compression

$A_s f_y = 0.85f'_c ba$, and therefore $a = A_s f_y/(0.85f'_c b)$

This is represented in the diagram below:

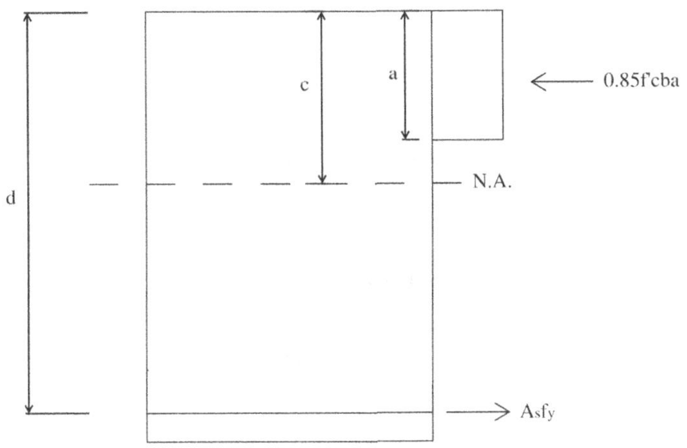
c
a
0.85f'cba
N.A.
d
Asfy

The minimum reinforcing in a concrete beam is the larger of the following two equations:

$3f'c^{0.5}b_w d/f_y$ or $200b_w d/f_y$

The maximum reinforcing does not have a simple equation but is a function of limiting the strain in the steel so that the mode of failure is not crushing of the concrete. This is done by setting the strain of steel to 0.005. Therefore:

$0.005 = ((d-c)/c)(0.003)$, solve for c

$a=\beta c$, for $f'c=4000$, $\beta=0.85$, for $f'_c > 4000$, $\beta = 0.85 - ((f'_c-4000)/1000)(0.05) > 0.65$

Then find the corresponding area of steel plugging into $A_s f_y=0.85f'_c ba$

Shear

The shear capacity of a concrete beam is the addition of the shear strength of the concrete and the reinforcing stirrups. Therefore:

$\phi V_n = \phi V_c + \phi V_s$, $\phi = 0.75$

$\phi V_c = 2\phi\sqrt{f'_c} b_w d$

$\phi V_s = \phi(A_v f_y d)/s$ Where

s = spacing of stirrups (in)
A_v = Area of vertical stirrups (in^2). Note: the cross section for shear often includes multiple vertical bars. A_v is the total area of all vertical legs

Spacing shall not be greater than = $A_v f_y/50b_w$

Two Way Shear

Use ACI section 22.6.5. The perimeter of the critical section (b_o) is determined by creating a rectangle a distance of d/2 from each face of the column.

The 2-way shear strength is the smallest of the following 3 equations:

$(2+4/\beta)\sqrt{f'_c} b_o d$ Where β = the ratio of the long side to the short side of the column

$(\alpha_s d/b_o+2) \sqrt{f'_c} b_o d$ Where α_s = 40 for interior columns, 30 for edge columns, and 20 for corner columns

$4\sqrt{f'_c}\, b_o d$

Axial

The axial strength of short columns not considering PΔ effects is:

Tied Columns

$\phi P_n = \phi 0.80(0.85f'_c(A_g - A_S) + f_yA_S)$
$\phi = 0.65$ for tied columns
A_g = gross area of column (in^2)

Spiral Columns

$\phi P_n = \phi 0.85(0.85f'_c(A_g - A_S) + f_yA_S)$
$\phi = 0.75$ for Spiral columns
A_g = gross area of column (in^2)

Considering eccentric effects, the use of interaction diagrams is needed:

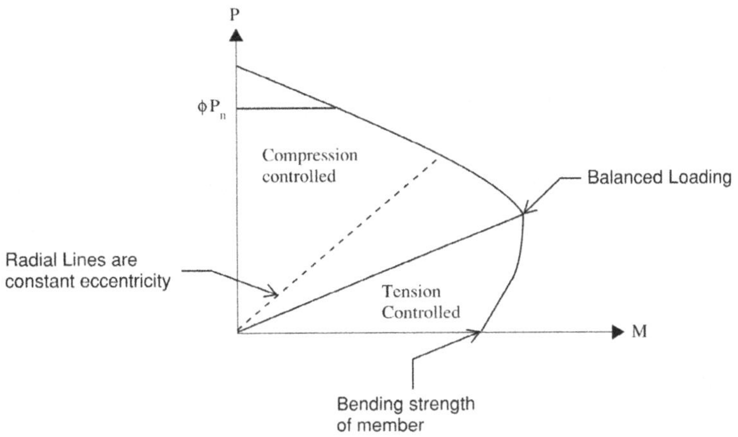
P
ϕPn
Compression controlled
Balanced Loading
Radial Lines are constant eccentricity
Tension Controlled
M
Bending strength of member

ACI also provides limits for the reinforcing of members in compression:

- Code Requirements for columns:
  - Minimum Longitudinal steel > $0.01A_g$
  - Maximum Longitudinal steel < $0.08A_g$
  - Minimum Number of Bars:
    - 4 for rectangular ties
    - 3 for triangular ties
    - 6 for spiral ties
  - Minimum size tie is #3 for #10 bars and smaller, #4 for #10 bars and larger
- Center to center tie spacing shall not be greater than:
  - 16(longitudinal bar diameter)
  - 48(tie diameter)
  - Least dimension of the column

Reinforcing Development and Details

For No. 6 and small bars with appropriate cover $l_d = \left(\frac{f_y \Psi_t \Psi_e}{25\sqrt{f'_c}}\right) d_b$

For No. 7 and larger bars with appropriate cover $l_d = \left(\frac{f_y \Psi_t \Psi_e}{25\sqrt{f'_c}}\right) d_b$

Ψ_t = 1.3 For horizontal bars which have more than 12 in of concrete cast below, otherwise 1.0
d_b = diameter of bar
Ψ_e = 1.5 for epoxy bars with cover less than $3d_b$ or clear spacing less than $6d_b$
= 1.2 for all other epoxy bars
= 1.0 for all other bars
The product of $\Psi_t\ \Psi_e$ shall not exceed 1.7

Development length may be reduced by A_s required/A_s provided

Development for standard hook $l_{dh} = (0.02\Psi_e f_y/\lambda\sqrt{f'_c})d_b > 6''$ or $8d_b$

λ = 1.0 for normal weight concrete and 0.75 for lightweight concrete

Hook development length shall also be reduced by Section 12.5.3

American Institute of Steel Construction (AISC)

Flexure

Flexural capacity of steel is determined by the length of lateral support. If a beam is fully laterally supported, the capacity is the plastic moment capacity:

The plastic moment is the yield strength times the plastic section modulus: $M_P = F_yZ$
The plastic section modulus is the summation of the areas in compression and tension multiplied by the distance from the center of gravity of each area to the plastic neutral axis.

For a symmetric W shape, $Z = 2(A_fY_f + A_w/2Y_w)$

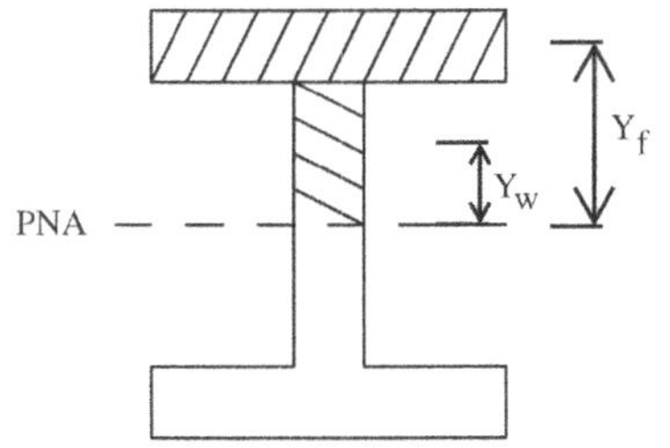
PNA
Y_w
Y_f

If a beam is not fully laterally supported, the capacity is a function of the unbraced length and will fail in lateral torsion buckling. For the purposes of this exam it is best to determine the capacity of a member from the design tables AISC 3-10. Find the intersection of the unbraced length and the applied load and then find the first dark line directly above this point to find the most efficient section.

Compression

The capacity of compression members are a function of the unbraced length. For non-doubly symmetric members there is a different capacity about each axis and it must be determined which is the controlling axis. To do this AISC Chapter 4 Tables provide the conversion factor to determine the equivalent unbraced length of the strong axis which is the ratio of the radii of gyration about each axis.

$(KL)_{Yequiv} = (KL)_X/(r_x/r_y)$

Then the design axial capacity can be determined from the appropriate table for each structural shape.

Tension

A steel tension member needs to be checked for 2 modes of failure. Yielding of the gross area and rupture of the effective net area

$\phi R_n = \phi F_y A_g$ Yielding $\phi = 0.9$
$\phi R_n = \phi F_u A_n$ Rupture. $\phi = 0.75$

The gross area includes no holes
The net area includes the holes. The diameter of the hole is obtained by adding 1/8" to the diameter of the bolt

Fatigue

Fatigue is covered in the specifications Appendix 3 of the AISC. The stress range is determined by the following equation:

$F_{SR} = (C_f/N)^{0.333} > F_{TH}$

N is the fluctuations for the life of the structure
C_f = Found from table A3.1
F_{TH} = Table A3.1

Bolt Strength

Bolted connections fail in shear or bearing.

For Shear:

$\phi R_n = \phi F_n A_b n$, $\phi = 0.75$

F_n = Shear strength of bolt from table J3.2. N is for threads included, X for threads excluded
n = number of bolts

Also note the capacity is multiplied by the number of shear planes on the bolt

For Bearing:

$\phi R_n = \phi 1.2 L_c F_u t < \phi 2.4 t F_u d$ $\phi = 0.75$

L_c = clear edge distance measured from edge of hole to edge of connected material
F_u = ultimate strength of connected material
t = thickness of connected material
d = bolt diameter

Block Shear

The main concept to understand for block shear is that a component of the capacity comes from tension which is the length of perimeter perpendicular to the load and the other comes in shear from the portion which is parallel to the tension. Then simply fill in the equation

$\phi R_n = \phi(0.6F_uA_{nv} + U_{bs}F_uA_{nt}) < 0.6F_yA_{gv} + U_{bs}F_uA_{nt}$, $\phi = 0.75$

U_{bs} = 1.0 if tension is uniform and 0.5 otherwise (Most often taken as 1.0)

A_{nv} = net area in shear (in^2)
A_{nt} = net area in tension (in^2)
A_{gv} = gross area in shear (in^2)

The diameter of the hole is 1/8" + diameter of the bolt

Weld Strength

The equation for the strength of a fillet weld is:

$\phi R_n = \phi 0.6F_{EXX}(0.707wL)$, $\phi = 0.75$

F_{EXX} = weld strength (ksi)
w = effective throat width (in)
L = length of weld (in)

However, this value is only for welds parallel to the force. The strength needs to be modified by the following equation to account for the angle of the weld to the load
$0.5\sin^{1.5}\theta + 1$, where
θ = angle of the weld to the force

National Design Standards for Wood Construction (NDS)

Note: Wood design for the PE exam uses Allowable Stress Design only

Wood Design consists of finding the allowable design stress for each mode of failure by multiplying the reference stress by the appropriate adjustment factors to find the design allowable stress. The breakdown for sawn lumber and structurally glued members is in Tables 4.3.1 and 5.3.1 respectively.

Sawn Lumber

All reference design values are found in the Supplement Tables 4A, 4B, 4C, 4D, and 4E
C_D = load duration factor as specified in Sec 2.3.2
C_M = moisture content factor = 1.0 if moisture content < 19%. If > 19% as specified in Tables 4A, 4B, 4C, 4D, and 4E
C_t = use table 2.3.3 temperature condition factor = 1.0 if sustained temperatures < 100. If > 100° use Sec 2.3.3
C_L = lateral support factor = 1.0 for fully laterally supported. Use Sec 3.3.3 otherwise
C_F = size factor = for 2" or 4" thick, use tables 4A and 4B. For 5" or thicker and exceeding 12" in depth = $(12/d)^{1/9} < 1.0$
C_{fu} = flat use factor for weak axis bending as specified in 4A, 4B, 4C, and 4F for 2" or 4" thick
C_i = incising factor. Refer to table 4.3.8
C_r = repetitive member factor use Sec 4.3.9
C_T = Buckling stiffness factor use Sec 4.4.2

$$C_p = \frac{1 + \frac{F_{cE}}{F*_c}}{2c} - \sqrt{\left(\frac{1 + \frac{F_{cE}}{F*_c}}{2c}\right)^2 - \frac{\frac{F_{cE}}{F*_c}}{c}}$$

$$F_{CE} = \frac{0.822E'_{min}}{\left({}^{l_e}/_{d}\right)^2}$$

c = 0.8 for sawn lumber

$F*_C = F_c$ multiplied by all factors except C_p

$E'_{min} = E_{min}C_MC_t$

Note: C_p needs to be evaluated about both the X-axis and Y-axis

only use the lesser of C_L or C_V

Structural Glued Laminated

All reference design values are found in the Supplement Tables 5A, 5B, 5C, and 5D

C_D = load duration factor as specified in section 2.3.2

C_M = moisture content factor = 1.0 if moisture content < 16%. If > 16% as specified in Sec 5.1.4.2

C_t = temperature condition factor = 1.0 if sustained temperatures < 100°. If > 100° use Sec 2.3.3

C_L = lateral support factor = 1.0 for fully laterally supported. Use Sec 3.3.3 otherwise

C_V = Volume factor = $(21/L)^{1/x}\,(12/d)^{1/x}(5.125/b)^{1/x} < 1.0$

L = length of bending member between points of zero moment (ft)

d = depth of bending member (in)

b = width of member (in)

x = 20 for southern pine, 10 for all other species

C_{fu} = weak axis bending factor as specified in Sec 5.3.7

C_C = curvature factor. For curved members = $1 - (200t/R)^2$

t = thickness of laminations

R = radius of curvature to inside face

C_I = taper factor. For tapered portions use Sec 5.3.9

C_{VR} = Shear Reduction factor. Use Sec 5.3.10

$$C_p = \frac{1+\frac{F_{cE}}{F*_c}}{2c} - \sqrt{\left(\frac{1+\frac{F_{cE}}{F*_c}}{2c}\right)^2 - \frac{\frac{F_{cE}}{F*_c}}{c}}$$

$$F_{CE} = \frac{0.822E'_{min}}{\left(l_e/d\right)^2}$$

c = 0.9 for Structural Glued Laminated

$F*_C$ = F_c multiplied by all factors except C_p

$E'_{min} = E_{min}C_MC_t$

Note: C_p needs to be evaluated about both the X-axis and Y-axis

only use the lesser of C_L or C_V

Connections

Connections are similar to other modes of failure in which the reference value is multiplied by adjustment factors to determine the design allowable stress. Dowel Type Fastener (bolts, lag screws, wood screws, nails, spikes, drift bolts, drift pins) stresses are determined from Table 11.3.1

For Lateral Loads:
$Z' = Z \times C_D\, C_M\, C_t\, C_g\, C_D\, C_{eg}\, C_{di}\, C_{tn}$

Withdrawal

$W' = W \times C_D C_M{}^2 C_t C_{eg} C_{tn}$

C_D = Duration Factor 11.3.2
C_M = Wet Service Factor 11.3.3
C_t = Temperature Factor 11.3.4
C_g = Group Action Factor 11.3.6
C_Δ = Geometry Factor 12.5.1
C_{eg} = End Grain Factor 12.5.2
C_{di} = Diaphragm Factor 12.5.3
C_{tn} = Toe Nail Factor 12.5.4

ACI 530 Specifications for Masonry Structures (MSJC)

Note: Masonry design for the PE Exam uses Allowable Stress Design

Bending

The service moment capacity is the lesser of the following 2 equations:

$M_m = F_b jkb_w d^2/2.$ Or $M_S = F_s j\rho b_w d^2$

F_s = Allowable stress of steel (32 ksi for grade 60 steel, 20 ksi for grade 40 or 50)
F_b = Allowable compressive stress = $f'_m 0.45$
ρ = Steel ratio = $A_s/b_w d$
$n = E_s/E_m$
$k = (2\rho n + (\rho n)^2)^{0.5} - \rho n$
$J = 1 - k/3$

Compression

Compression members are differentiated by reinforced or unreinforced. Either way first determine the appropriate equation by calculating the slenderness ratio = h/r

$$r = \sqrt{I/A}$$

For square cross sections, this reduces to r = 0.289b

Then determine the appropriate equation by determining the ratio h/r. For reinforced:

$P_a = (0.25f'_m A_n + 0.65A_{st}F_s)(1 - (h/(140r)^2)$ for $h/r < 99$ Equation 8-21

$P_a = (0.25f'_m A_n + 0.65A_{st}F_s)(70r/h)^2)$ for $h/r > 99$ Equation 8-22

For unreinforced:

For $h/r \leq 99$

$$f_a \leq F_a 1/4 f'_m \left[1 - \left(\frac{h}{140r}\right)^2\right]$$ Equation 8-16

For $h/r > 99$

$$f_a \leq F_a 1/4 f'_m \left(\frac{70r}{h}\right)^2$$ Equation 8-17

Precast/Prestressed Concrete Institute (PCI)

Prestressing Stresses

In addition to the stress from external forces, prestressed beams are subject to the stresses from the strands. There are 2 types of stress from the applied Prestressing force (P):

1. Compression stress due to the strands: P/A
2. Bending due to the eccentricity of the strands: Pe(c)/I

When calculating total stress, be aware of signs. The eccentric prestress force causes a negative moment which will offset any positive bending.

Prestressing Flexure

- Moment capacity is similar to reinforced concrete however the prestressing introduces a new resistance component.

$$M_n = A_{ps}f_{ps}\left(d_p - \frac{a}{2}\right) + A_sf_y\left(d - \frac{a}{2}\right)$$
$$Since\ T = C,\ 0.85f_c'ba = A_{ps}f_{ps} + A_sf_y$$

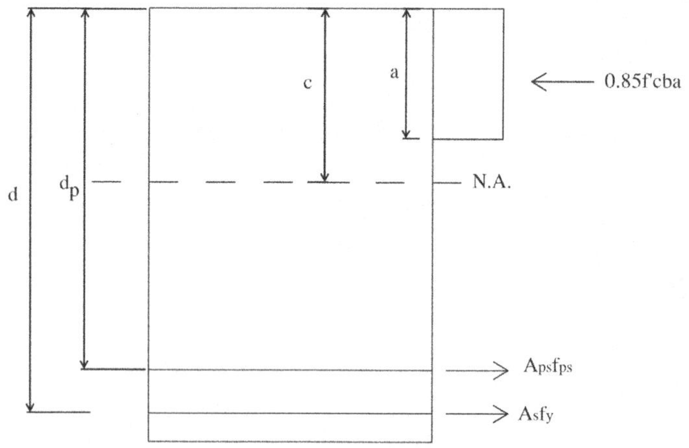
c
a
0.85f'cba
d
dp
N.A.
Apsfps
Asfy

Shipping/Handling

- Precast/prestressed members need to be transported. This introduces stresses which may be different from the in-place design conditions.
- PCI provides provisions for the handling of precast members to limit cracking.
- The modulus of rupture of the section must be greater than applied stress due handling.
- Modulus of Rupture, $f_r = 7.5\sqrt{f'_c}$
- PCI provides equations for the moments during lifting of typical pick point configurations. This can be found in Figure 8.3.1.
- The situation in which lifting or transportation occurs requires an additional multiplier as outlined in Table 8.3.1.

OSHA 1910 General Industry and 1926 Construction Safety Standards

Some main concepts include:

- Excavation Safety: Except for excavations in rock, anything deeper than 5 ft must be stabilized to prevent cave-in. This may be achieved by providing appropriate earth retention systems or sloping at appropriate rates. This is determined by the depth of excavation, soil type, and other requirements. 1926 Subpart P – Excavations

- Fall protection: Drop-offs must be protected from fall based on the height of the drop. Some examples of protection include temporary fences, nets, or lifelines. 1910 Subpart D – Walking Working Surfaces and 1926 Subpart M – Fall Protection

- Power Line Hazards: For power lines which are electrified, all construction activities must be a minimum distance from the lines. This is based on the voltage of the lines. Typically, the safe operational distance is 10 ft. for lines less than 50 kV and typically 35 ft. for lines greater than 50 kV. 1926 Subpart V – Electric Power Transmission and Distribution

- Confined Spaces: Anyone required to enter confined spaces must be appropriately trained and equipped. Oxygen must be monitored and kept at an acceptable level. Subpart AA – Confined Spaces

- Personal Protective Equipment (PPE): Equipment required by any personnel present on a job site. Examples are acceptable head protection and steel toed shoes. 1910 Subpart I – Personal Protective Equipment

International Building Code (IBC)

Loads and Load Combinations

Load combinations as per IBC are:

Section 1605.2

- $1.4(D + F)$ (Equation 16-1)
- $1.2(D + F) + 1.6(L + H) + 0.5(L_r \text{ or } S \text{ or } R)$ (Equation 16-2)
- $1.2(D + F) + 1.6(L_r \text{ or } S \text{ or } R) + 1.6H + (f_1L \text{ or } 0.5W)$ (Equation 16-3)
- $1.2(D + F) + 1.0W + f_1L + 1.6H + 0.5(L_r \text{ or } S \text{ or } R)$ (Equation 16-4)
- $1.2(D + F) + 1.0E + f_1L + 1.6H + f_2S$ (Equation 16-5)
- $0.9D + 1.0W + 1.6H$ (Equation 16-6)
- $0.9(D + F) + 1.0E + 1.6H$ (Equation 16-7)

- $f_1 = 1$ for places of public assembly live loads in excess of 100 pounds per square foot (4.79 kN/m2), and parking garages; and 0.5 for other live loads.
- $f_2 = 0.7$ for roof configurations (such as saw tooth) that do not shed snow off the structure, and 0.2 for other roof configurations.

In addition, the following exceptions apply:

1. Where other factored load combinations are specifically required by other provisions of this code, such combinations shall take precedence.
2. Where the effect of H resists the primary variable load effect, a load factor of 0.9 shall be included with H where H is permanent and H shall be set to zero for all other conditions.

Fire Ratings

- Fire resistance rating – Minimum amount of time a building element or component can continue to perform its structural function.

Special Inspections

- Special Inspections – Inspections of construction materials and components by qualified professionals for compliance with codes and construction documents
- Can be continuous (Full time) or periodic (Part time or intermittent)

American Welding Society (AWS)

Welding Symbols and Types

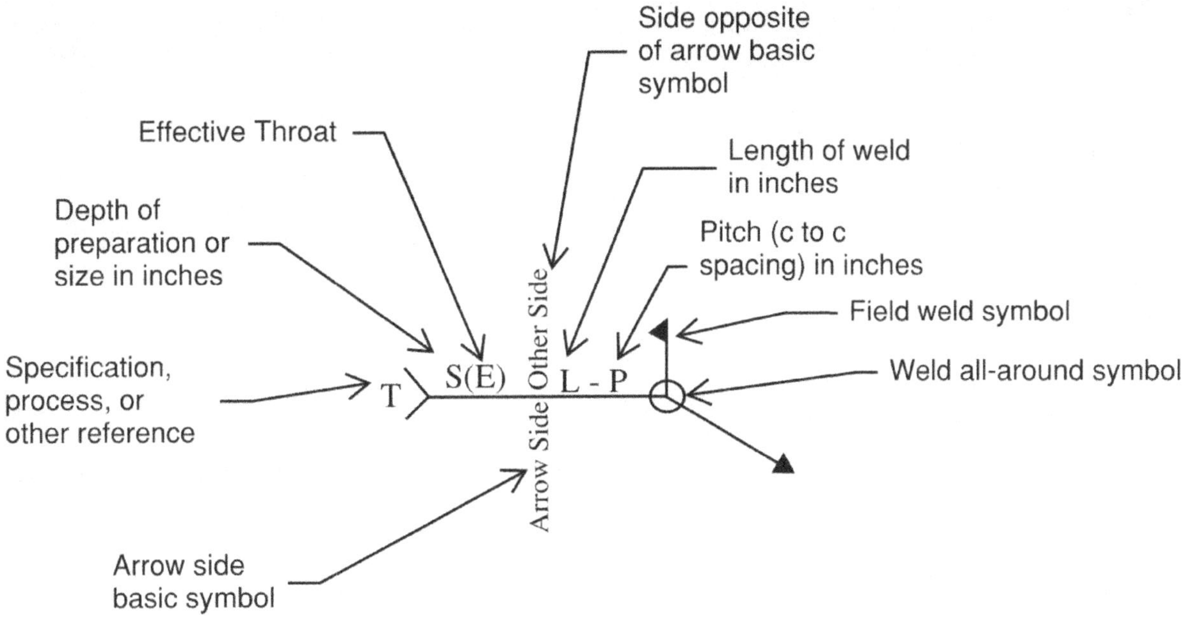
Side opposite of arrow basic symbol
Effective Throat
Length of weld in inches
Depth of preparation or size in inches
Pitch (c to c spacing) in inches
Field weld symbol
Specification, process, or other reference
T
S(E)
Other Side
L - P
Weld all-around symbol
Arrow Side
Arrow side basic symbol

Back	Fillet	Plug or Slot	Groove or Butt						
			Square	V	Bevel	U	J	Flare V	Flare Bevel

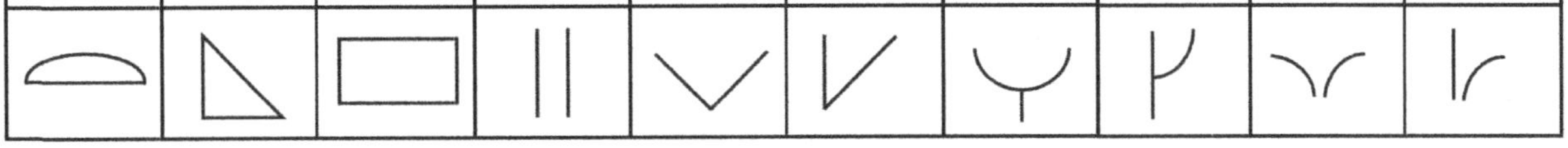

- Back Weld – A second weld created on the back side of a joint after the initial weld is made on the opposite side.
- Fillet Weld – Joining two pieces of metal either perpendicular or at an angle to each other.
- Plug Weld – Connecting two members by welding around and with a hole in one of the members.
- Groove Weld – Weld that connects 2 members parallel to each other. It is a weld that consists of an opening between 2 surfaces.

Advanced Statics

3D Statics

Statics in 3 dimensions introduces additional equations of equilibrium due to the third axis. Apply the same basic principles:

$\sum F_X = 0$
$\sum F_Y = 0$
$\sum F_Z = 0$
$\sum M_X = 0$
$\sum M_Y = 0$
$\sum M_Z = 0$

- First determine location of origin (0, 0, 0)
- Determine X, Y, and Z component of all forces
- Determine moment from each component about each axis
- Moment about an axis is the perpendicular distance from a force component to that axis
- Forces parallel to an axis has zero moment about that axis
- Forces that run through an axis have zero moment about the axis

Moving Loads

- Moving Loads are most often from Live Load due to traffic
- Need to analyze position of load to cause the greatest stress
- Shear in general is greatest when loads are at the support
- Positive moment in general is greatest with the loads at midspan
- Negative Moment is greatest with the load close to the support

Hinges

Hinges are supports at which there is a zero moment and only an axial and vertical force can be transferred

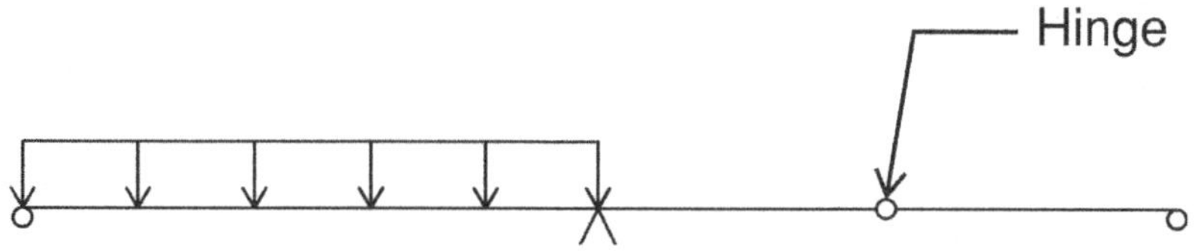
Hinge

Hinges are best analyzed by taking free body diagrams to either side of the hinge

Cables

- Cables carry load only in tension
- Acts as axial two force tension members
- Can be analyzed similarly to trusses use the method of joints

Consider the example below:

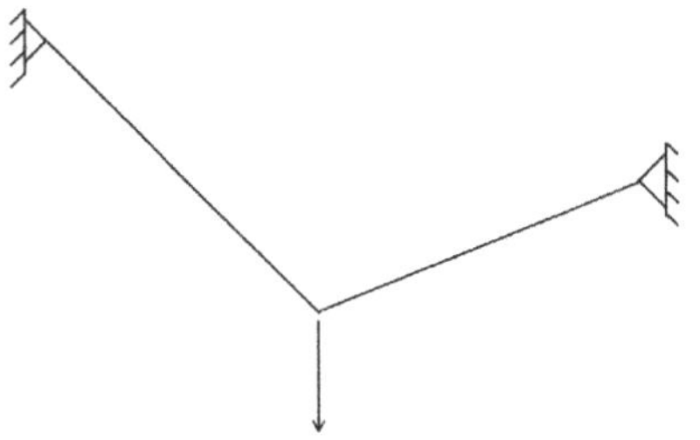

Determine support reactions by drawing free body diagram of entire system

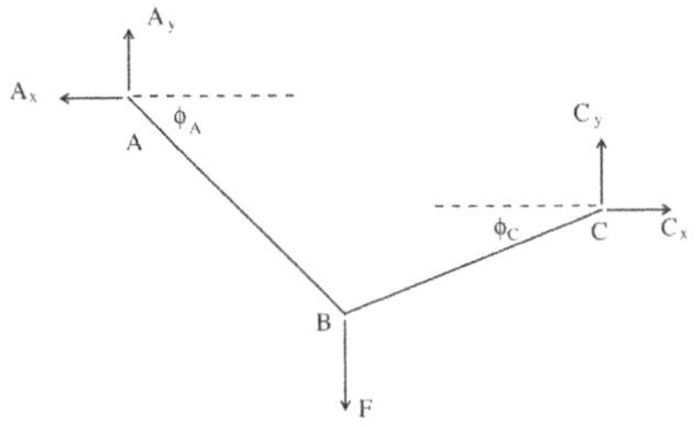

A_y
A_x
ϕ_A
A
C_y
ϕ_C
C
C_x
B
F

Then you can take free body diagrams of individual points to determine axial tensions:

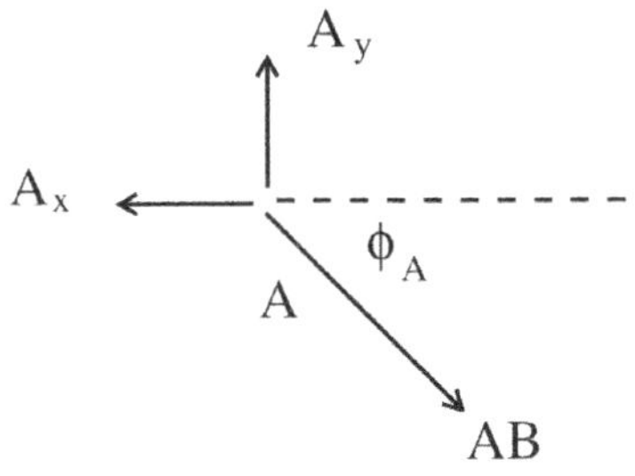

A_y
A_x
ϕ_A
A
AB

Miscellaneous Topics

Eccentrically Loaded Bolts

The bolt which will see the largest resultant force in a bolt group is the upper most bolt on the side of which the force is applied

The resultant $R = \sqrt{(V_P + V_e)^2 + {H_e}^2}$

V_p = vertical force per bolt = P/N
P = applied force (kips)
N = number of bolts

V_e = Force due to eccentricity = Pex_c/I_o
X_c = distance in X direction from centroid of bolt group to the subject bolt

The moment of inertia the sum of the distances of all bolts in the x and y directions if all bolts have the same diameter. Sum the distances of each bolt to the appropriate neutral axis:

$I_o = \Sigma y^2 + \Sigma x^2$ (in^4/in^2)

$H_e = Pey_c/I_o$

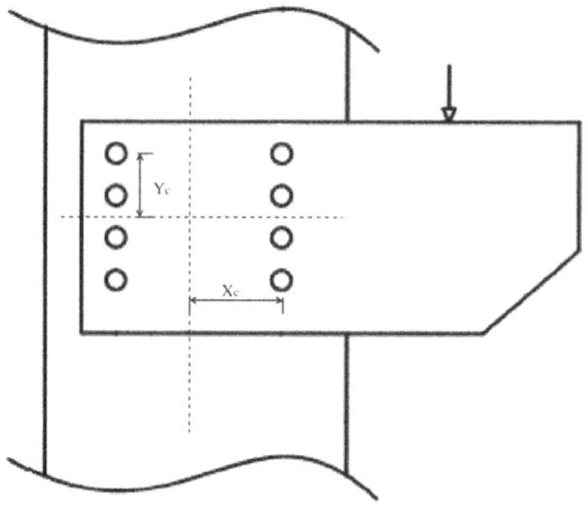

Yc
Xc

Elongation/Contraction Due to Axial Loading

The elongation or contraction of a beam due to an applied force is determined from the following equation:

$\Delta = PL/AE$
P = applied force (kips)
L = length of beam (in)
A = cross sectional area (in^2)
E = modulus of elasticity (ksi)

Thermal Expansion/Contraction

The thermal elongation or contraction for steel is determined by the following equation:

$\Delta = L_o\alpha(t_1 - t_o)$
L_o = original length of beam (in)
α = coefficient of thermal expansion
t_1 = new temperature (degrees Fahrenheit)
t_o = original temperature (degrees Fahrenheit)

Composite Material Beams

For composite members, the stresses in the beams can only be determined by analysis of similar materials. To do this, the stronger material should be converted to an equivalent area of the weaker material. This is done using the ratio of the moduli of elasticity $n = E_1/E_2$

Then calculate the equivalent width of the weaker material $b_{equiv} = bn$

The member can then be analyzed as a single material.

Torsional Shear

The torsional shear stress of a circular member $\tau = Tr/J$,
T = torsional moment
r = radius of member
J = polar moment of inertia
For a hollow circular member $J = \pi/2(r_o^4 - r_i^4)$

Surcharge Loads

Increased pressure on a buried structure due to a point load is determined from Boussinesq's equation

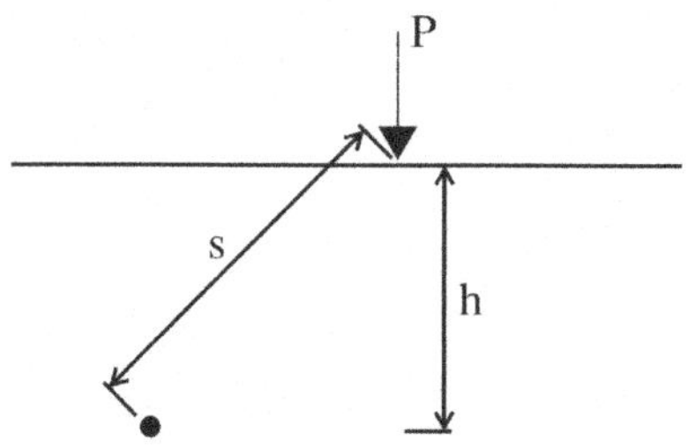
P
s
h

$$\Delta p_v = \frac{3h^3P}{2\pi s^5}$$

h = Distance below surface (ft)
P = Surcharge point load (lb)
s = Inclined distance from load to buried structure (ft)

Deflections

- Not all deflections are covered under beam chart equations
- Moment Area Method can be used for the determination of angle and deflection
- First draw the moment diagram
- Determine the M/EI diagram by dividing points on the moment diagram by EI
- The angle between tangents of two points is the area of the M/EI diagram between points
- The deflection of a point from the tangent of another is area of the M/EI times the distance from the centroid of this area to the desired point of deflection

Horizontal Shear Stress

Horizontal shear stress

$\tau = VQ/Ib$

V = Applied Shear Force (kips)
Q = First Moment of the Desired Area = ay.
a = Cross Sectional Area from Point of Desired Shear Stress to Extreme Fiber (in^2)
y = Distance from Centroid of Beam to Centroid of Area "a" (in)
I = Moment of Inertia of Beam (in^3)
b = Width of Member (in)

MORNING BREADTH PRACTICE EXAM

Question M1

The chart below gives an estimation of the area of sub-grade to be cut at stations along the baseline of a roadway. Determine most nearly the total volume of excavation in cubic yards from the data.

STATION	**AREA OF CROSS SECTION (FT^2)**
1+00	0
1+50	155
2+00	170
2+50	65
3+00	0

(A) 722
(B) 19500
(C) 800
(D) 19222

Question M2

Given the dimensions of the 8′ tall proposed concrete retaining wall shown below determine the area of formwork in square feet required to complete construction of the wall portion only after the footing has been poured.

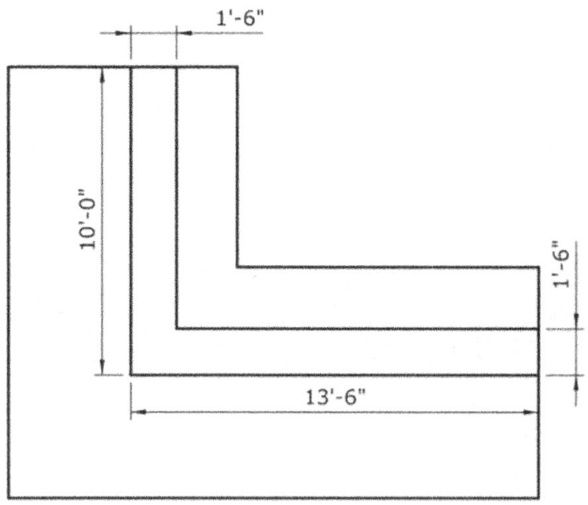

1'-6"
10'-0"
1'-6"
13'-6"

(A) 950
(B) 400
(C) 376
(D) 450

Question M3

Given the scheduling data shown below, Identify all of the tasks on the critical path.

Activity	Predecessor	Duration
A		5 (Months)
B	A	3
C	A	2
D	C	3
E	D	4
F	E	1
G	D	2
H	G	4

(A) A
(B) B
(C) C
(D) D
(E) E
(F) F
(G) G
(H) H

Question M4

Design alternatives are being proposed for a bridge replacement with a life span of 30 years. Alternative A has an initial cost of $100,000. It will also require anticipated maintenance costs of 10,000 and 15,000 at years 10 and 20 respectively. Use a rate of 3% and determine most nearly the present worth of the design alternative.

(A) $99050
(B) $110444
(C) $113250
(D) $115746

Question M5

A soil sample has 20% fines and over 50% finer than the no. 4 sieve. It also has a Liquid Limit of 55 and a Plastic Limit of 23. Determine the USCS classification group symbol of this soil.

(A) SC
(B) GC
(C) CL
(D) SM

Question M6

Given the beam and loading conditions shown below, determine most nearly the maximum service moment (kip-ft).

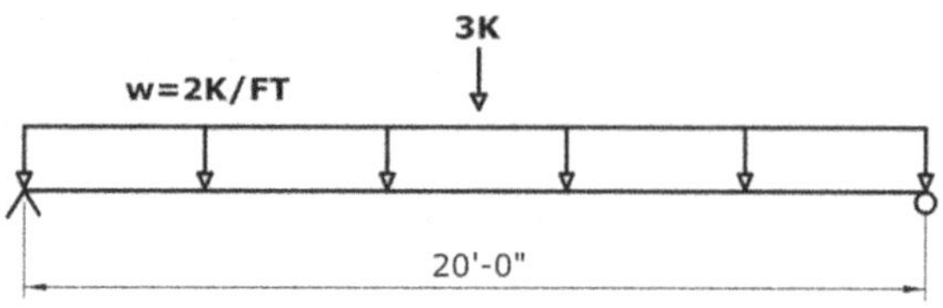
3K
w=2K/FT
20'-0"

(A) 100
(B) 115
(C) 9.58
(D) 57.5

Question M7

Assign the appropriate theoretical effective length factor for the columns with end conditions as shown:

0.5

0.7

1.0

2.0

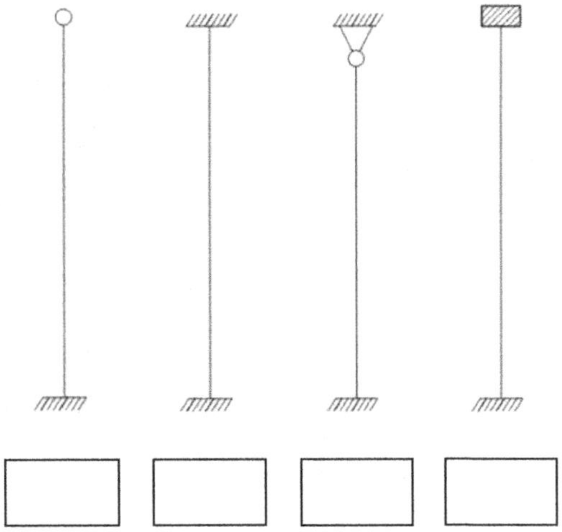

Question M8

Which of the following accurately represents the shear diagram for the beam and loading conditions shown?

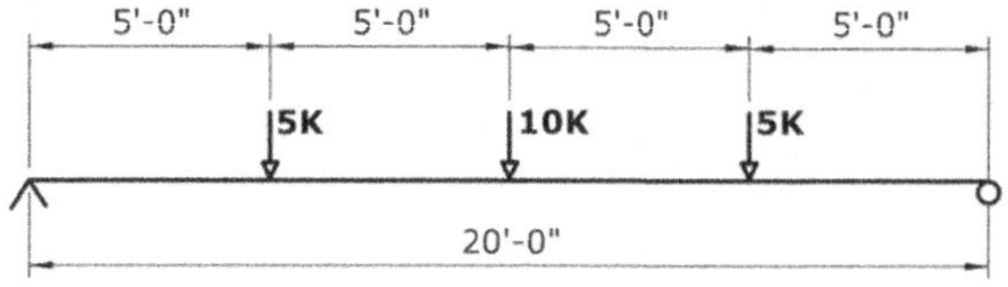

5'-0"
5'-0"
5'-0"
5'-0"
5K
10K
5K
20'-0"

(A)

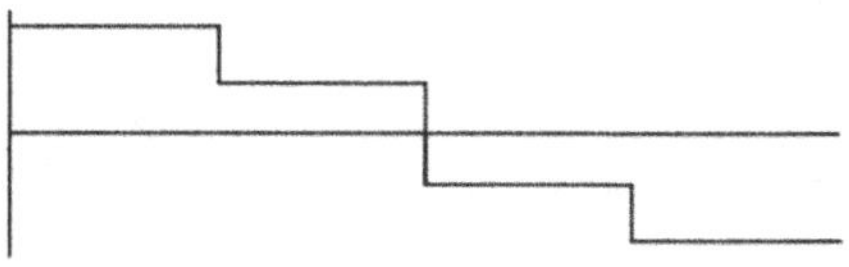

(B)

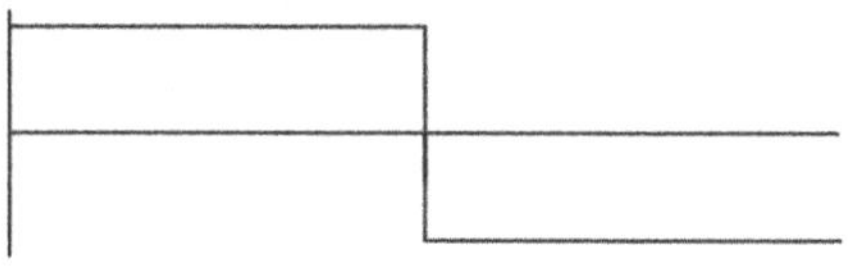

(C)

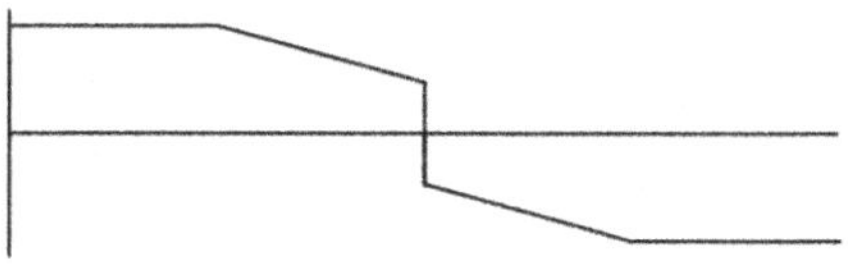

(D)

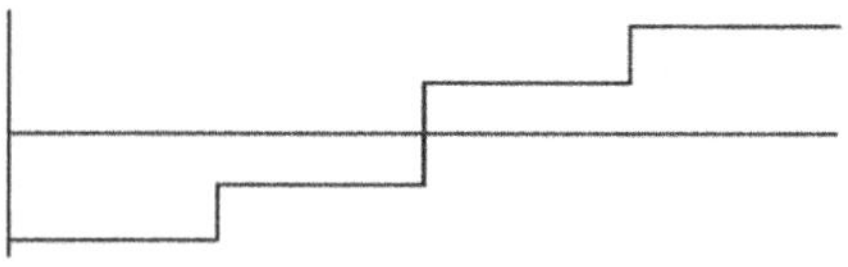

Question M9

Given the beam and loading condition shown below, determine most nearly the moment at point C in Kip-ft.

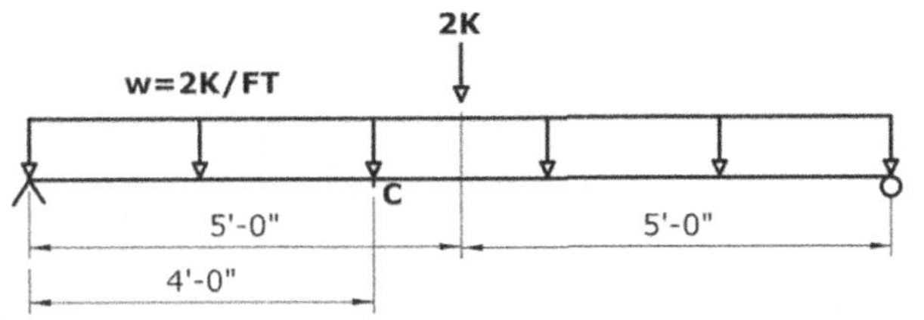
2K
w=2K/FT
C
5'-0"
5'-0"
4'-0"

(A) -10
(B) 15
(C) 28
(D) 40

Question M10

Three 6-inch diameter x 12 inch long concrete cylinders were taken from a concrete batch used to form a retaining wall. The cylinders broke at compressive loads of 102000, 111000, and 100500 pounds. Determine most nearly the average compressive strength of the concrete (psi).

(A) 3200
(B) 4000
(C) 3600
(D) 3700

Question M11

A horizontal curve has a point of curvature at station 1+50 and a point of tangency at station 2+25. If the interior angle of the curve is 10°15′, determine most nearly the radius of the curve.

(A) 330′
(B) 420′
(C) 500′
(D) 600′

Question M12

A crest vertical curve has $G_1 = 2\%$, $G_2 = -4\%$ and a PVI elevation of 155.0′ at station 10+00. A bridge has a low point elevation of 170′ to the bottom of the girder at station 8+00. Determine most nearly the vertical clearance between the road and low point of the bridge if the length of curve is 800′.

(A) 17.0′
(B) 20.5′
(C) 26.0′
(D) 29.5′

Question M13

A vertical Curve has $G_1=-3\%$ and $G_2=2\%$. The PVI elevation is 150.0′ at station 10+00. The Length of Curve is 600′. The station of the low point is ______________.

Question M14

A horizontal curve is proposed to be constructed around a building. A car traveling along the curve has a line of sight past the building to an object in the road from the beginning of curve to the end of curve. Determine most nearly the minimum distance from the center of road to the building for a required stopping sight distance of 220′ to the object if the radius of the curve is 700′.

(A) 8.63′
(B) 15.55′
(C) 20.20′
(D) 35.0′

Question M15

Backfill material is to be transported to a construction site to achieve a proposed grade behind a retaining wall. A total of 100 cubic yards is necessary in the final condition. Determine most nearly in cubic yards how much fill is needed to be taken from the offsite location if there is a swell factor of 1.07, an assumed loss during transport of 5%, and the backfill will be compacted to 90%?

(A) 100
(B) 102
(C) 106
(D) 109

Question M16

A mix design for a parking garage will use 500 lbs of cement per cubic yard. What volume of water in cubic feet per cubic yard should be used to achieve a w/c ratio of 0.65?

(A) 5.21
(B) 6.00
(C) 4.31
(D) 7.50

Question M17

A retaining wall without a batter is 15.5' tall from the bottom of footing and is used to hold granular soils with an angle of internal friction of 25. Determine most nearly the resultant service moment at the base of the footing using Rankine active earth pressure from the soil only. The angle of fill is horizontal. Neglect friction between the wall and the soil and use a soil density of 115 pcf.

(A) 24.7 k-ft
(B) 26.5 k-ft
(C) 28.9 k-ft
(D) 32.1 k-ft

Question M18

A saturated soil sample has a weight of 50 lbs. The sample is placed in an oven and then weighed to measure 42 lbs. The specific gravity of the soil was determined to be 2.4. Determine most nearly the void ratio of the sample.

(A) 0.28
(B) 0.40
(C) 0.42
(D) 0.46

Question M19

A vehicle is traveling at a velocity of 100 ft/s on a 2% incline. If the driver has a 2 second breaking perception reaction time, determine the total distance in feet it takes to stop the vehicle. Assume a coefficient of friction of 0.3.

(A) 200.0
(B) 684.5
(C) 480.9
(D) 700.0

Question M20

Which of the following is added to the tangent runout to calculate the superelevation transition distance?

(A) Superelevation runoff
(B) Cross slope
(C) Approach Grade
(D) Degree of Curvature

Question M21

Assuming fully turbulent flow, determine most nearly the velocity in ft/sec for a pipe with a diameter of 2 ft, a length of 500 ft and a head loss due to friction of 3 ft. Use a Darcy friction factor 0.02.

(A) 1.0
(B) 5.0
(C) 6.2
(D) 11.4

Question M22

A rectangular open channel is 6 ft wide and has a water depth of 3 ft. Determine most nearly the flow rate (cfs). The channel has a roughness coefficient of 0.015 and a slope of 0.003.

(A) 130
(B) 138
(C) 145
(D) 200

Question M23

For a head loss due to friction of 40 ft, determine most nearly the volumetric flow rate in gallons per min for a pipe with length 100′ and diameter of 0.25 ft. Use a Hazen-Williams coefficient of 140.

(A) 200
(B) 265
(C) 410
(D) 432

Question M24

Which of the following defines the time base for a stream hydrograph?

(A) Time from the base flow until the peak flow
(B) Time from the peak flow until flow drops below the base flow
(C) Time that the flow exceeds the base flow
(D) Time that the flow drops below the base flow

Question M25

A circular culvert is to be designed to handle the runoff from two areas of drainage. The first has an area of 15 Acres, a runoff coefficient of 0.18, and a rainfall intensity of 1.5 in/hr. The second has an area of 10 Acres, a runoff coefficient of 0.22, and a rainfall intensity of 1.5 in/hr as well. Determine most nearly the minimum culvert area in square feet to limit the flow velocity to 0.26 ft/s.

(A) 28.28
(B) 29.50
(C) 31.0
(D) 62.0

Question M26

Given the truss configuration shown below, identify the 0-Force members. The truss is simply supported and points A and H. (point and click example)

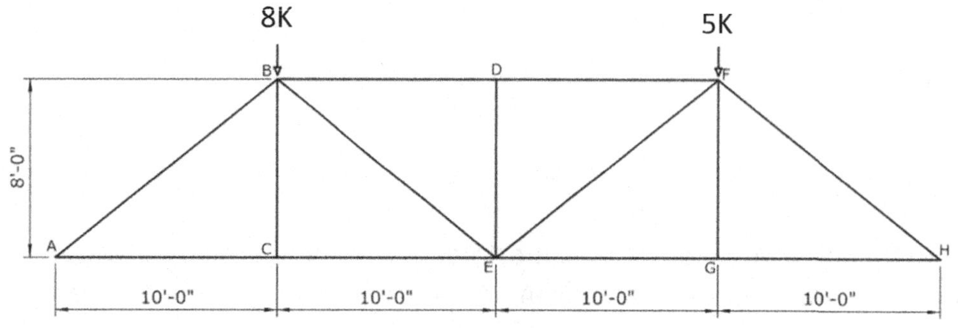
8K
5K
A
B
C
D
E
F
G
H
8'-0"
10'-0"
10'-0"
10'-0"
10'-0"

Question M27

Given the truss configuration shown below, what is the support reaction at node A? Neglect the self-weight of the truss. The truss is simply supported and points A and H.

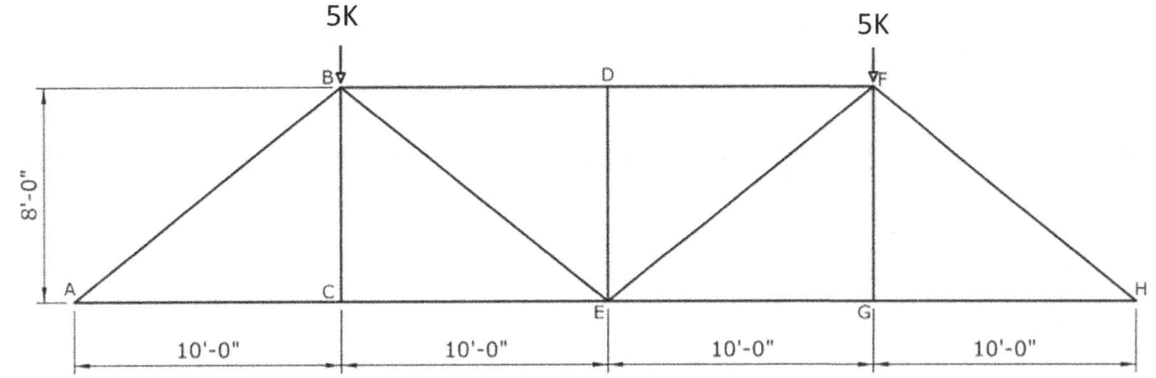
5K
5K
B
D
F
8'-0"
A
C
E
G
H
10'-0"
10'-0"
10'-0"
10'-0"

(A) 2 Kips
(B) 5 Kips
(C) 6.67 Kips
(D) 10 Kips

Question M28

Given the truss and loading condition in question M27, what is the axial force in member C-E? Neglect the self-weight of the truss.

(A) 0 Kips
(B) 5 Kips
(C) 6.24 Kips
(D) 8.71 Kips

Question M29

Which of the following is most often the cause of rebar corrosion?

(A) Chloride intrusion of concrete
(B) Freeze-thaw action
(C) Delayed Ettringite Formation
(D) High water to cement ratio

Question M30

Which of the following mix design properties has the greatest impact on the strength of the concrete?

(A) w/c ratio
(B) Percent Coarse Aggregate
(C) Void Ratio
(D) Percent Fine Aggregate

Question M31

Water exits a reservoir through a pipe 50′ below the water surface. Determine the velocity of flow at the exit of the pipe. Assume frictionless flow and the discharge is at atmospheric pressure.

(A) 45.66 ft/s
(B) 56.75 ft/s
(C) 60.10 ft/s
(D) 75.55 ft/s

Question M32

Which of the following is not an assumption of the Bernoulli energy conservation equation?

(A) The fluid is incompressible
(B) There is no fluid friction
(C) Changes in thermal energy are negligible
(D) The potential energy is zero

Question M33

Flow from pipes A and B connect to exit out a single pipe C and have velocities of 1.2 and 0.8 ft/s respectively. If pipe A has an area of 3.0 ft^2 and pipe B an area of 4.0 ft^2, what is the flow rate for pipe C in cubic feet per second?

(A) 3.6
(B) 4.0
(C) 6.0
(D) 6.8

Question M34

After an onsite investigation, the soil profile below has been developed from borings. Determine most nearly the effective stress at a depth of 40 ft for the saturated soil densities indicated.

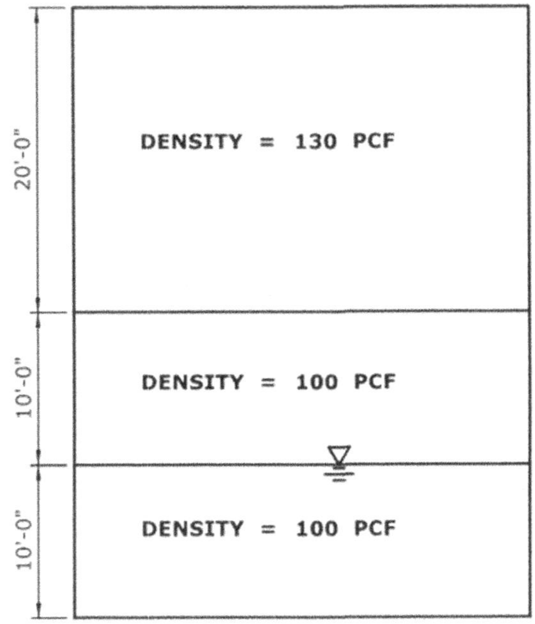

20'-0"
DENSITY = 130 PCF
10'-0"
DENSITY = 100 PCF
10'-0"
DENSITY = 100 PCF

(A) 2600 psf
(B) 3550 psf
(C) 3975 psf
(D) 4600 psf

Question M35

A simply supported 4" wide X 6" deep beam is 15' long and has two point loads of 10 kips applied at 5' and 10' from the left end, what is the maximum bending stress of the cross section?

(A) 25 ksi
(B) 35 ksi
(C) 50 ksi
(D) 55.5 ksi

Question M36

A retaining wall is to be designed by limiting the shear stress in the stem to 0.1 ksi/ft. As shown below the horizontal force due to earth pressure is calculated as 9 Kips. The retaining wall has a batter of 8:1. Determine most nearly the minimum width of the base of the stem, b, if the depth to the flexural reinforcement is 6".

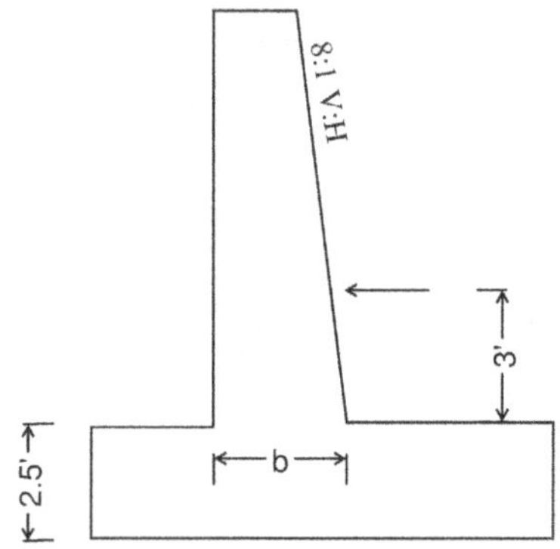
8:1 V:H
3'
b
2.5'

(A) 6.0"
(B) 7.5"
(C) 8.25"
(D) 12"

Question M37

Which of the following roadside safety barriers is most appropriate for an object which is 2'-6" perpendicular from the edge of roadway?

(A) Concrete Barrier
(B) 3-Cable Guiderail System
(C) Metal Beam Rail
(D) None

Question M38

For the traffic counts shown below, determine the peak hourly traffic volume.

Time	Volume
8:00-8:15	500
8:15-8:30	560
8:30-8:45	650
8:45-9:00	625
9:00-9:15	630
9:15-9:30	600
9:30-9:45	540
9:45-10:00	460

(A) 2260
(B) 2455
(C) 2505
(D) 2600

Question M39

Given a 6″ wide X 12″ deep cantilever beam with a point load at the free end of 1 K, determine the maximum deflection on the beam for a length of 10′. Use a modulus of elasticity of 3605 ksi.

(A) 0.10″
(B) 0.18″
(C) 0.28″
(D) 0.50″

Question M40

Determine the specific gravity of a soil sample. The total volume is 2 cubic ft and the volume of the soil is 1.5 cubic ft. The degree a saturation is 75% and the moisture content is 0.1.

(A) 2.2
(B) 2.5
(C) 2.8
(D) 2.95

STRUCTURAL DEPTH PRACTICE EXAMS

Question S1

An unheated open-air parking garage is fully exposed and in a terrain category B and has a risk category of II. Determine most nearly the flat roof snow load. Use a ground snow load of 35 psf.

Design Code: ASCE 7 – *Minimum Design Loads for Buildings and Other Structures, 2010, 3rd Printing*

(A) 20.0 psf
(B) 22.2 psf
(C) 26.5 psf
(D) 35.0 psf

Question S2

An unheated open-air structure fully exposed is to be built in a terrain category C and a risk category of I. The roof has a slope of 20 degrees and consists of asphalt shingles. Determine the design roof snow load using a ground snow load of 40 psf. Ignore unbalanced loading effects.

Design Code: ASCE 7 – *Minimum Design Loads for Buildings and Other Structures, 2010, 3rd Printing*

(A) 22.5
(B) 24
(C) 30
(D) 40

Question S3

Find the seismic base shear for ordinary precast shear walls using the equivalent lateral force procedure. Use an effective seismic weight of 200 K and a design spectural response acceleration parameter at short periods of 0.3. The structure is a risk category of II.

Design Code: ASCE 7 – *Minimum Design Loads for Buildings and Other Structures, 2010, 3rd Printing*

(A) 10 K
(B) 20 K
(C) 45 K
(D) 100 K

Question S4

Determine the design spectural response acceleration parameter at short periods for an area of very dense soil and soft rock. Use a Mapped Spectural Response Acceleration parameter of 0.5.

Design Code: ASCE 7 – *Minimum Design Loads for Buildings and Other Structures, 2010, 3rd Printing*

(A) 0.33
(B) 0.38
(C) 0.40
(D) 0.55

Question S5

Determine the appropriate multiple presence factor for the analysis of a 5 Girder Concrete Bulb Tee bridge for 3 lanes loaded.

Design Code: AASHTO – *LRFD Bridge Design Specifications, 7th Ed.*

(A) 0.85
(B) 0.90
(C) 1.00
(D) 1.20

Question S6

For which of the following scenarios is the equivalent lateral force procedure for calculating seismic base shear not permitted?

Design Code: ASCE 7 – *Minimum Design Loads for Buildings and Other Structures, 2010, 3rd Printing*

(A) Category B, Light Frame Construction
(B) Category F, Light Frame Construction
(C) Category D, Structure height=150' and no structural irregularities, and $T < 3.5T_S$
(D) Category D, Structure height=200' and no structural irregularities, and $T > 3.5T_S$

Question S7

The appropriate seismic design category for risk category of II, an S_1=0.065, and a S_{ds}=0.25 is ___________.

Design Code: ASCE 7 – *Minimum Design Loads for Buildings and Other Structures, 2010, 3rd Printing*

Question S8

For a hospital 4th floor used for operating, the design distributed dead load is 30 psf. Determine the total factored distributed load. Do not consider reduction of Live Load.

Design Code: ASCE 7 – *Minimum Design Loads for Buildings and Other Structures, 2010, 3rd Printing*

(A) 132 psf
(B) 90 psf
(C) 36 psf
(D) 150 psf

Question S9

For an interior column A, determine the design live load for the 3rd floor. The occupancy is for school classrooms. The column has a tributary area of 300 ft^2 and a K_{LL}=4.

Design Code: ASCE 7 – *Minimum Design Loads for Buildings and Other Structures, 2010, 3rd Printing*

(A) 27 psf
(B) 35 psf
(C) 40 psf
(D) 50 psf

Question S10

In the figure below, determine the maximum factored vertical reaction at the left support. Do not include any live load factor reductions.

Design Code: ASCE 7 – *Minimum Design Loads for Buildings and Other Structures, 2010, 3rd Printing*

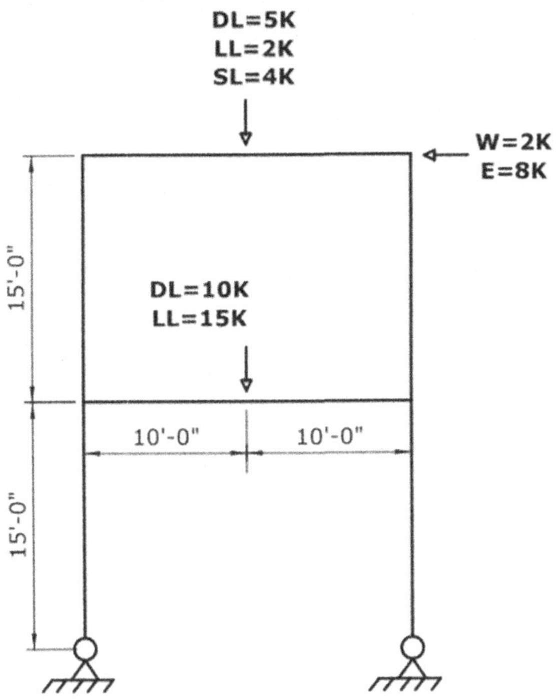
DL=5K
LL=2K
SL=4K
W=2K
E=8K
DL=10K
LL=15K
10'-0"
10'-0"
15'-0"
15'-0"

(A) 10.0 K
(B) 18.5 K
(C) 25.2 K
(D) 28.9 K

Question S11

Which of the following is not a required test for fillet welds?

Design Code: AWS – *American Welding Society (AWS D1.1, D1.2, and D1.4)*

(A) Macrotech
(B) All-Weld Metal Tension
(C) Side Bend
(D) Reduced Section Tension

Question S12

A concrete pile is used to support a platform. The pile is 50 feet long and is embedded into stiff soil for 40'. The remaining 10 ft is fully exposed. What length of the pile can be considered as laterally supported?

Design Code: IBC – *International Building Code, 2015 Ed.*

(A) 35'
(B) 40'
(C) 45'
(D) 50'

Question S13

Determine the fire-resistance rating requirement of an interior bearing wall for a building with a Type I construction class with no external protection.

Design Code: IBC – *International Building Code, 2015 Ed.*

(A) 0 hr
(B) 1 hr
(C) 2 hrs
(D) 3 hrs

Question S14

Determine the AASHTO Live Load distribution factor for moment of an interior beam considering one design lane loaded. The cross section of the bridge consists of steel girders spaced 6' with an 8" composite concrete deck. The bridge span is 60' and the longitudinal stiffness parameter is 800600.

Design Code: AASHTO – *LRFD Bridge Design Specifications, 7th Ed.*

(A) 0.35
(B) 0.45
(C) 0.62
(D) 1.12

Question S15

Determine most nearly the design axial strength of the short compression member as shown in the cross section below with 6-#9 bars. Use f'_c = 4000 psi and Grade 60 Reinforcing. The confinement steel consists of ties.

Design Code: ACI 318- *Building Code Requirements for Structural Concrete 2014*

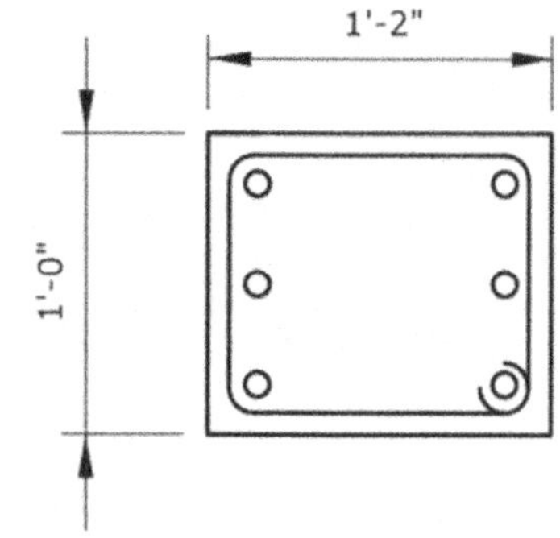

1'-2"
1'-0"

(A) 455.0 K
(B) 473.6 K
(C) 491.0 K
(D) 520.0 K

Question S16

A concrete haunch has been designed to support a prestressed beam. Determine to the nearest whole inch, the length L to fully development the main flexural reinforcement for a standard 90 degree hook of a #6 bar. Use Grade 60 non epoxy reinforcement and $f'_c = 4000$ psi. The required steel is equal to that provided.

Design Code: ACI 318- *Building Code Requirements for Structural Concrete 2014*

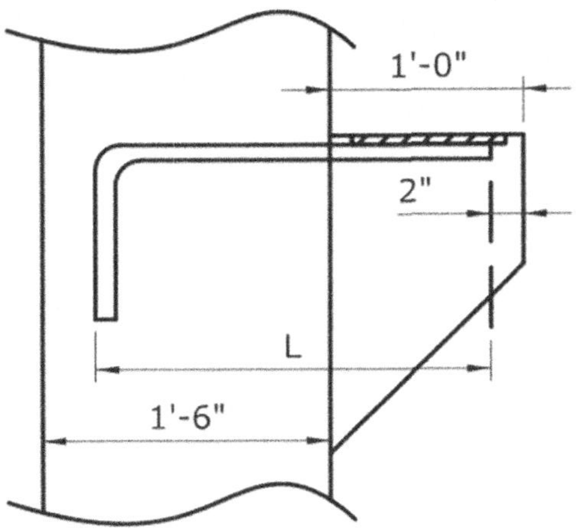
1'-0"
2"
L
1'-6"

(A) 10"
(B) 14"
(C) 19"
(D) 20"

Question S17

Determine the minimum height of a simply supported non-prestressed beam with a span length of 50′. The reinforcing is Grade 40.

Design Code: ACI 318- *Building Code Requirements for Structural Concrete 2014*

(A) 30″
(B) 36″
(C) 37.5″
(D) 42″

Question S18

1′-0″ wide concrete T-Beams support a 6″ slab and are spaced at 11′ on center. Determine most nearly the effective flange width for flexure design of an interior beam. The span length is 25′.

Design Code: ACI 318- *Building Code Requirements for Structural Concrete 2014*

(A) 87″
(B) 100″
(C) 120″
(D) 132″

Question S19

Determine most nearly the maximum and minimum reinforcing limits in square inches for a 12″ wide beam with a 15″ depth to reinforcing. Use f'_c = 4000 psi and Grade 60 reinforcement.

Design Code: ACI 318- *Building Code Requirements for Structural Concrete 2014*

(A) 0.6 and 3.25
(B) 0.57 and 3.25
(C) 0.6 and 2.5
(D) 1.0 and 3.50

Question S20

A uniform distributed load is applied to the frame with prismatic members which is shown below. Match the appropriate approximation equation with its correct location.

Design Code: ACI 318- *Building Code Requirements for Structural Concrete 2014*

$w_u l_n^2/10$

$w_u l_n^2/11$

$w_u l_n^2/14$

$w_u l_n^2/16$

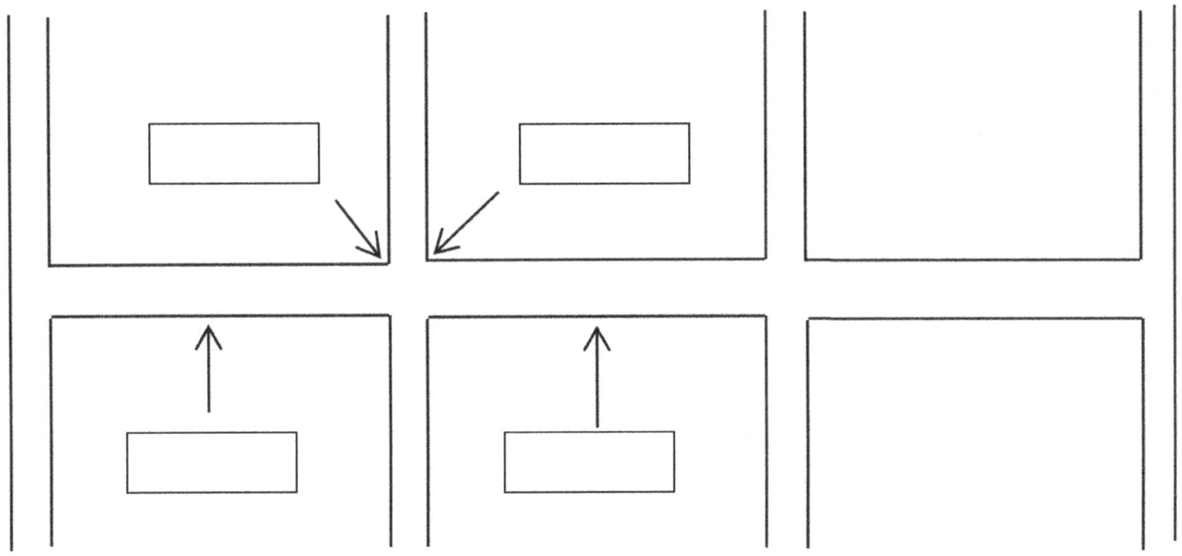

Question S21

Identify all of the types of Prestress losses from the following:

(A) Elastic shortening of concrete
(B) Shrinkage of concrete
(C) Creep of concrete
(D) Jacking force of tendons
(E) Friction
(F) Relaxation of steel
(G) Member proportions

Question S22

For a 12" wide beam with a depth of reinforcing of 24", determine the minimum spacing of #4 U-shaped stirrups for shear resistance of a 70 kip shear force. Use f'_c = 4000 psi and Grade 60 reinforcing.

Design Code: ACI 318- *Building Code Requirements for Structural Concrete 2014*

(A) 10.1"
(B) 12.25"
(C) 14.55"
(D) 18.21"

Question S23

The structural shape below is a compact beam with a yield strength of 50 ksi and a web thickness of ½". Determine most nearly the plastic moment.

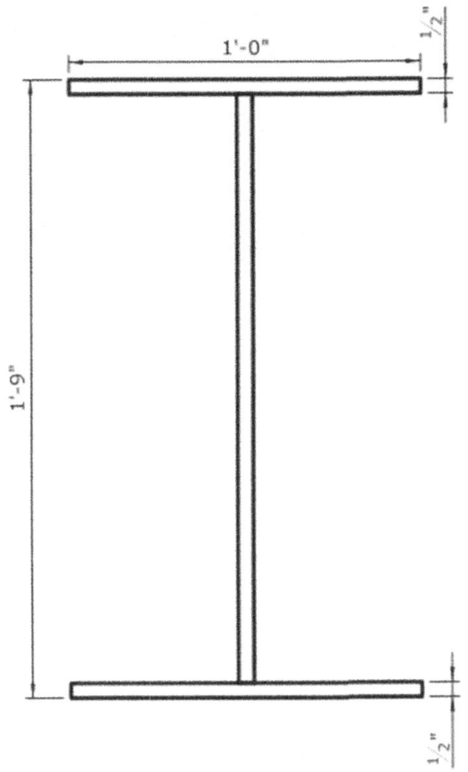
1'-0"
1'-9"
1/2"
1/2"

(A) 720 ft-k
(B) 900 ft-k
(C) 1200 ft-k
(D) 8000 ft-k

Question S24

A bearing type mechanical connection is shown below. The connection will be made with 7/8" diameter ASTM A325X bolts. Determine the minimum number of bolts needed to resist the factored shear loading of 200 kips.

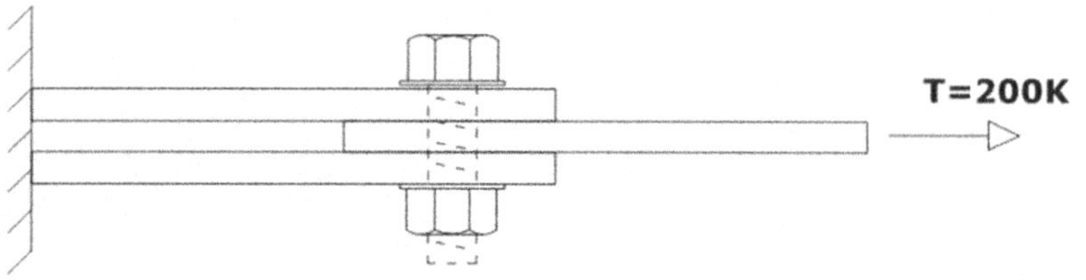
T=200K

Design Code: AISC – *Steel Construction Manual, 14th Ed.*

(A) 3
(B) 4
(C) 6
(D) 10

Question S25

A ½" thick tension member is connected to a gusset plate on a truss. Given the bolt configuration and dimensions as shown, determine the block shear capacity of the section for A36 steel.

Design Code: AISC – *Steel Construction Manual, 14th Ed.*

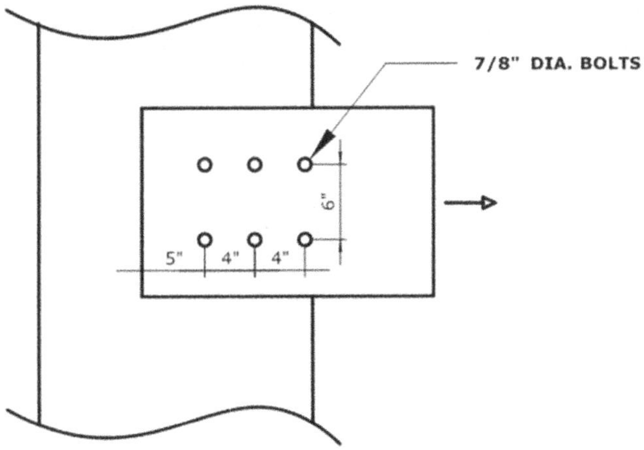
7/8" DIA. BOLTS
6"
5"
4"
4"

(A) 267 K
(B) 300 K
(C) 319 K
(D) 423 K

Question S26

Determine the design strength for a 7/8" fillet weld in tension as shown below. Use an E70 electrode.

Design Code: AISC – *Steel Construction Manual, 14th Ed.*

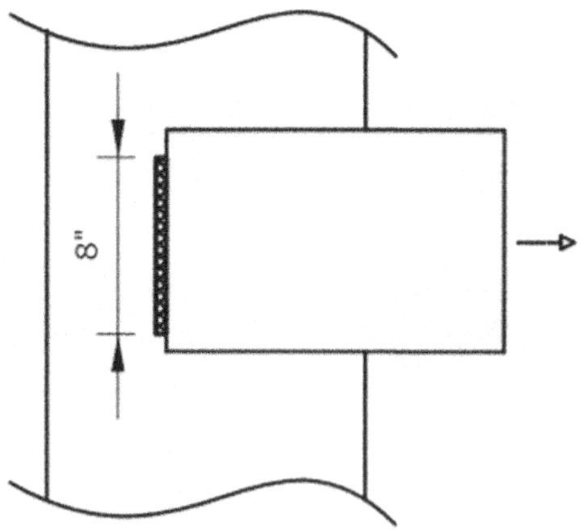
8"

(A) 120 K
(B) 156 K
(C) 200 K
(D) 234 K

Question S27

A 12" x 6" Glued Laminated 26F-V2 SP/SP column is 20' long and is subjected to an axial load parallel to the grain. It is braced about its weak axis at mid-height and the moisture content is less than 16%. Determine the design allowable compressive stress. The controlling load case is Dead + Live + Snow.

Design Code: NDS – *National Design Specifications for wood construction, 2015 Ed.*

(A) 1553 psi
(B) 1850 psi
(C) 2250 psi
(D) 2400 psi

Question S28

Determine the minimum number of 0.177" diameter threaded nails embedded 2" in Red Pine to resist 500 lbs in tension. The controlling load is dead + live under normal temperature conditions and the moisture content is 10%.

Design Code: NDS – *National Design Specifications for wood construction, 2015 Ed.*

(A) 4
(B) 6
(C) 8
(D) 12

Question S29

The bolted connection below is subjected to a 50 Kip eccentric force in the plane of the faying surface. Determine the resultant force on upper most bolt in the right column.

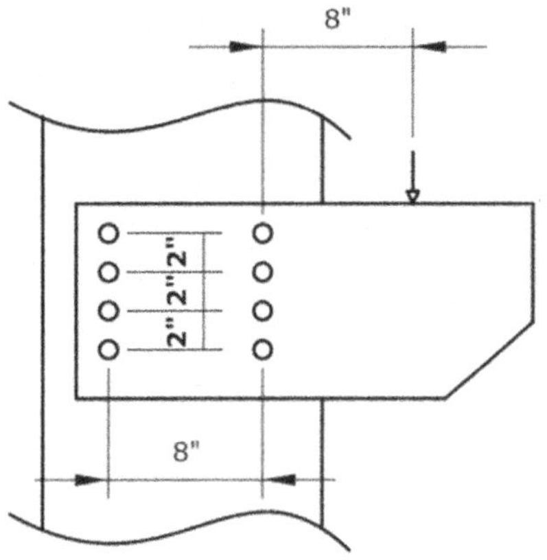
8"
2"
2"
2"
8"

.

(A) 23.16 K
(B) 27.50 K
(C) 33.55 K
(D) 45.0 K

Question S30

In wood design, the "24" in the combination symbol 24F-V4 indicates which of the following properties?

Design Code: NDS – *National Design Specifications for wood construction, 2015 Ed.*

(A) The depth of the member is 24"
(B) The length of the member is 24'
(C) The reference allowable bending stress is 2400 lb/in^2
(D) The allowable tension is 2400 b/in^2

Question S31

Given the continuous post-tensioned pier cap and loading conditions shown below, which strand profile is most appropriate?

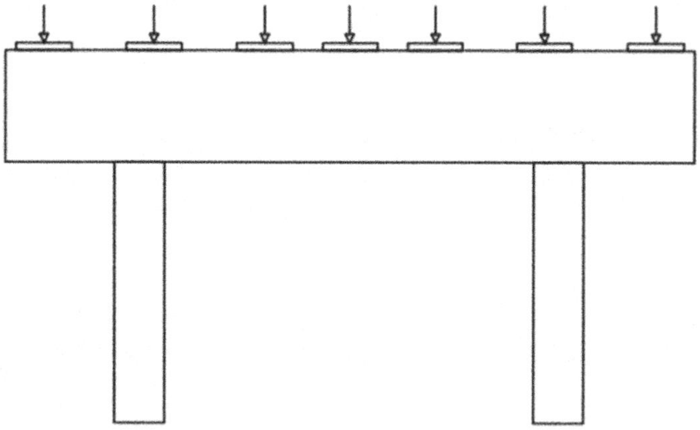

(A)

(B)

(C)

(D)

Question S32

An 8" X 10" beam has a depth to the prestressing tendons of 8". Given the loading condition shown below determine most nearly the maximum bending stress at the extreme tension face. Neglect the self-weight of the beam.

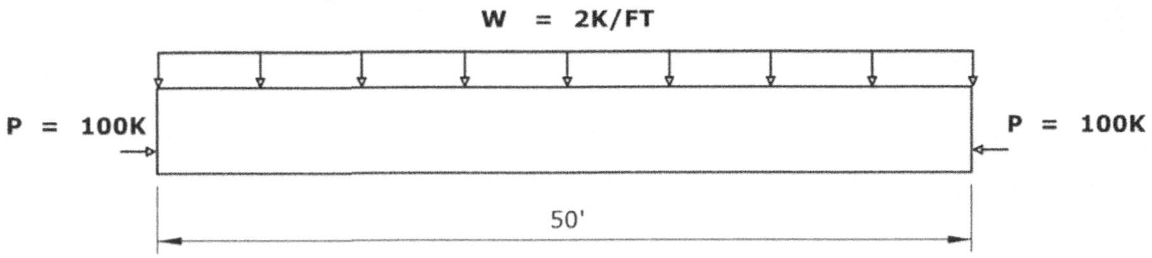

W = 2K/FT
P = 100K
P = 100K
50'

(A) 51.25 ksi
(B) 52.75 ksi
(C) 54.0 ksi
(D) 56.25 ksi

Question S33

A non-prestressed precast wall panel is designed to be 6" deep by 6' wide by 40' long. It will be transported and handled in the yard using a standard 4-point pick. Determine most nearly the minimum concrete compressive strength for no discernible cracking of the beam during yard handling only.

Design Code: PCI – *PCI Design Handbook: Precast and Prestressed Concrete 7th Ed. 2010*

(A) 2630 psi
(B) 3350 psi
(C) 4040 psi
(D) 4500 psi

Question S34

A steel girder for a bridge is 120 feet long. Determine the design contraction due to thermal movement. The bridge was set at a temperature of 65 and is designed using a high and low temperature of 95 and 10 degrees Fahrenheit. Use a thermal coefficient for steel of 0.0000065.

(A) 0.25″
(B) 0.5″
(C) 0.65″
(D) 0.75″

Question S35

A concrete beam is reinforced at the tension face with a ½″ thick steel plate as shown. Determine the distance from the neutral axis to the tension face. Use E_s = 29000 ksi and E_c = 3605 ksi.

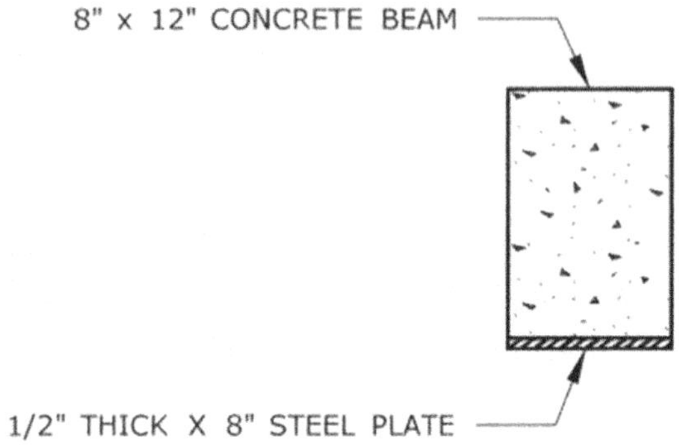

8" x 12" CONCRETE BEAM
1/2" THICK X 8" STEEL PLATE

(A) 4.9″
(B) 5.5″
(C) 6.25″
(D) 7.1″

Question S36

The beam shown below has 3 lines of reinforcement. Determine the effective depth of the flexural reinforcement.

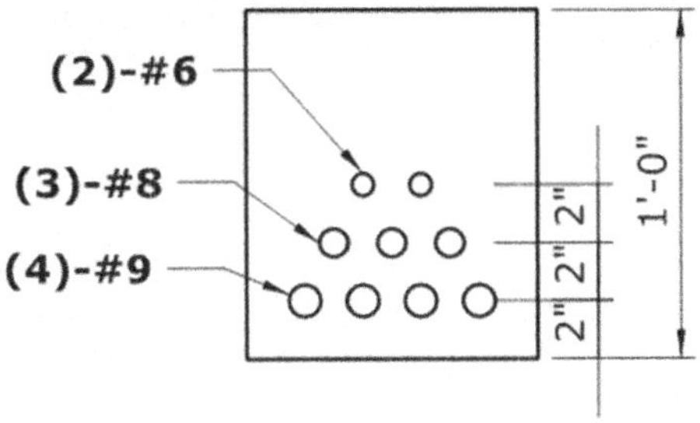

(2)-#6
(3)-#8
(4)-#9
2"
2"
2"
1'-0"

(A) 3.14"
(B) 8.0"
(C) 8.5"
(D) 8.86"

Question S37

Choose the appropriate symbol for a field ¼" fillet weld with a length of 4" as shown (point and click example):

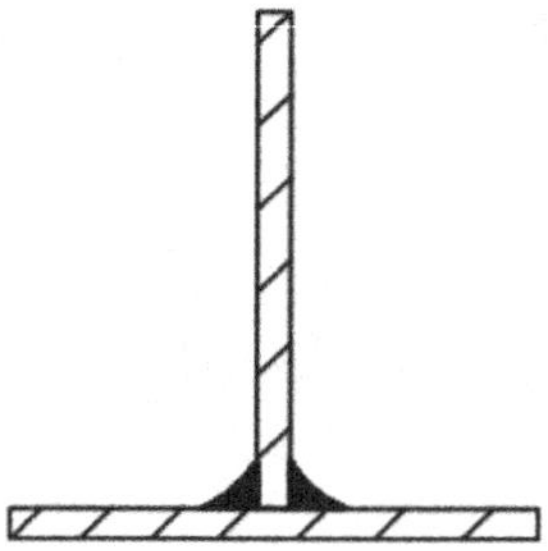

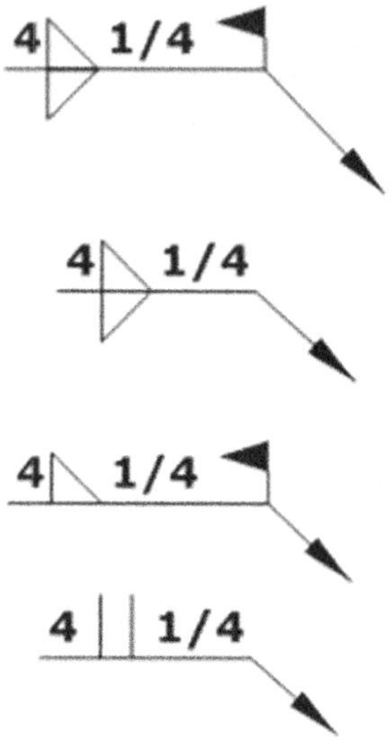
4 1/4
4 1/4
4 1/4
4 1/4

Question S38

Which of the following statements is the main reason for limiting the maximum amount of steel in a concrete beam?

(A) Ensures a reasonable speed of production
(B) Prevents concrete cover to reinforcing steel being minimized
(C) Ensures a ductile failure of the beam
(D) Ensures the steel achieves ultimate strength

Question S39

Which of the following is a prohibited weld type?

Design Code: AWS – *American Welding Society (AWS D1.1, D1.2, and D1.4)*

(A) Fillet welds less than 3/16"
(B) V-Groove welds in flat position
(C) Full penetration groove welds
(D) Overhead fillet welds

Question S40

According to OSHA standards determine the minimum horizontal distance from the edge of the working surface to the outer edge of a safety net for a drop off of 12'.

Design Code: OSHA CFR 29 – *Occupational Safety and Health Standards (Part 1910 and 1926)*

(A) 8'
(B) 10'
(C) 12'
(D) 13'

Question S41

For a five-span continuous bridge choose which spans to add a distributed load will cause the maximum load effect for positive flexure in the third span (Point and click example).

Question S42

According to OSHA safety standards determine the minimum height of a standard railing.

Design Code: OSHA CFR 29 – *Occupational Safety and Health Standards (Part 1910 and 1926)*

(A) 36”
(B) 42”
(C) 48”
(D) 52”

Question S43

For a circular sign support as shown below, determine the unfactored torsional shear stress caused by the resultant wind load of 5 kips.

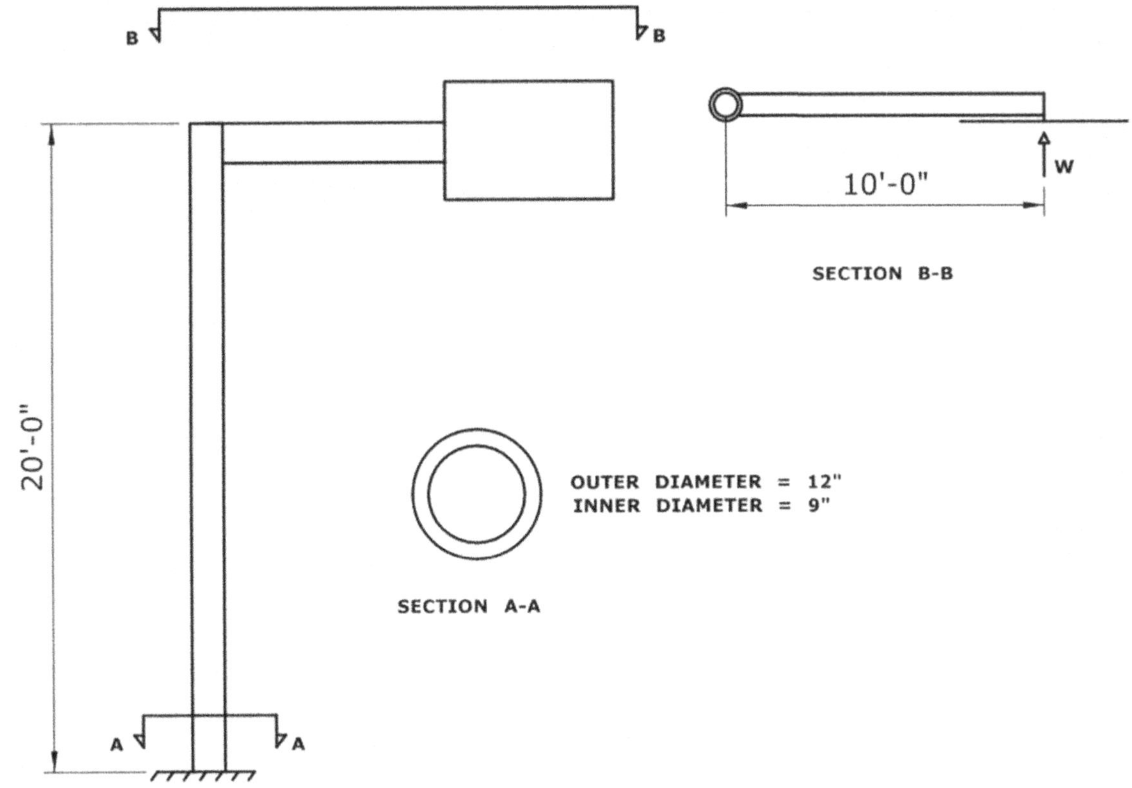
B
B
20'-0"
A
A
10'-0"
W
SECTION B-B
OUTER DIAMETER = 12"
INNER DIAMETER = 9"
SECTION A-A

(A) 2.6 ksi
(B) 4.4 ksi
(C) 5.6 ksi
(D) 10.5 ksi

Question S44

A glued laminated wood beam is shown below. Determine the horizontal shear stress at the interface of panels 1 and 2 for a shear force of 50 K.

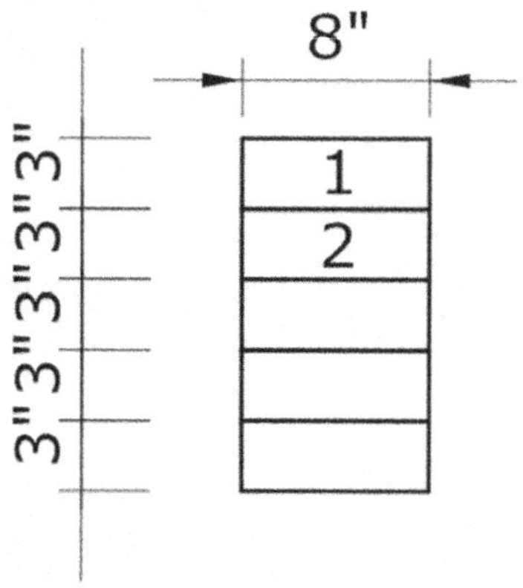
8"
1
2
3"
3"
3"
3"
3"

(A) 400 psi
(B) 650 psi
(C) 940 psi
(D) 1050 psi

Question S45

For the description of the concrete exposure listed in the table below match the correct minimum w/c ratio and concrete f'c:

Design Code: ACI 318- *Building Code Requirements for Structural Concrete 2014*

0.40	3500 psi
0.50	4000 psi
0.55	5000 psi

	w/c	f'c (psi)
Concrete in contact with water and low permeability is required		
Concrete exposed to moisture and an external source of chlorides		
Concrete exposed to freezing and thawing cycles with limited exposure to water		

Question S46

In the figure below determine the additional snow load in lbs per foot due to drift. Use a ground snow load of 40 psf and a p_s = 25 psf.

Design Code: ASCE 7 – *Minimum Design Loads for Buildings and Other Structures, 2010, 3rd Printing*

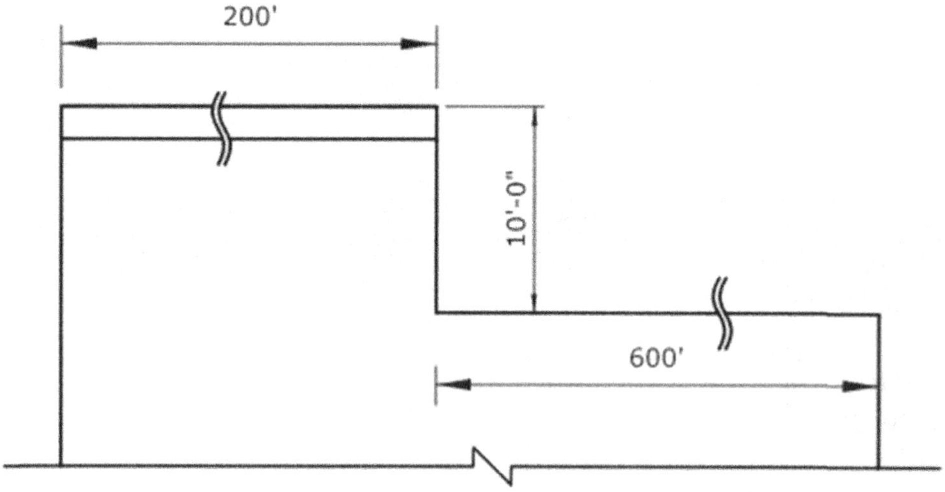

200'
10'-0"
600'

(A) 900
(B) 1294
(C) 1500
(D) 2200

Question S47

Determine the unfactored effective weight of a column supporting 2 stories as shown below. The first floor is used for storage. Do not account for Live Load reductions of the column.

Design Code: ASCE 7 – *Minimum Design Loads for Buildings and Other Structures, 2010, 3rd Printing*

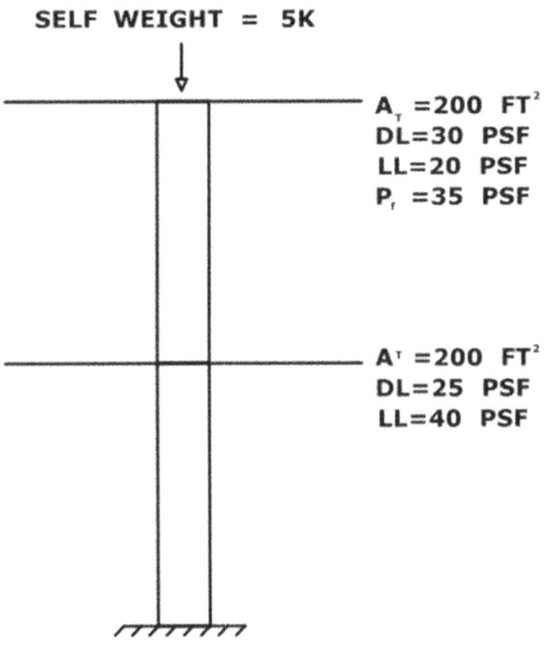
SELF WEIGHT = 5K
A_T =200 FT^2
DL=30 PSF
LL=20 PSF
P_r =35 PSF
A_T =200 FT^2
DL=25 PSF
LL=40 PSF

(A) 19.4 K
(B) 23.5 K
(C) 28.0 K
(D) 35.0 K

Question S48

Determine the minimum design lateral force for the 2nd story of a building with a seismic design category of A. The total dead load of story 2 is 150 K.

Design Code: ASCE 7 – *Minimum Design Loads for Buildings and Other Structures, 2010, 3rd Printing*

(A) 0.0 K
(B) 1.5 K
(C) 20.5 K
(D) 75.0 K

Question S49

The minimum 28-Day concrete compressive strength for micro piles is ____________.

Design Code: IBC – *International Building Code, 2015 Ed.*

Question S50

Which of the following is the minimum concrete reinforcing cover for a precast prestressed pile exposed to seawater?

Design Code: IBC – *International Building Code, 2015 Ed.*

(A) 2”
(B) 2.5”
(C) 3”
(D) 4”

Question S51

Which AASHTO limit state is used for yielding of steel structures?

Design Code: AASHTO – *LRFD Bridge Design Specifications, 7th Ed.*

(A) Strength I
(B) Service I
(C) Service II
(D) Extreme I

Question S52

A simply supported beam is to be designed to support the loading from a truck driving across the length of the span. The truck has a front axle load of 8 kips and a rear axle load of 14 kips. Determine most nearly the maximum service shear force for design of the beam.

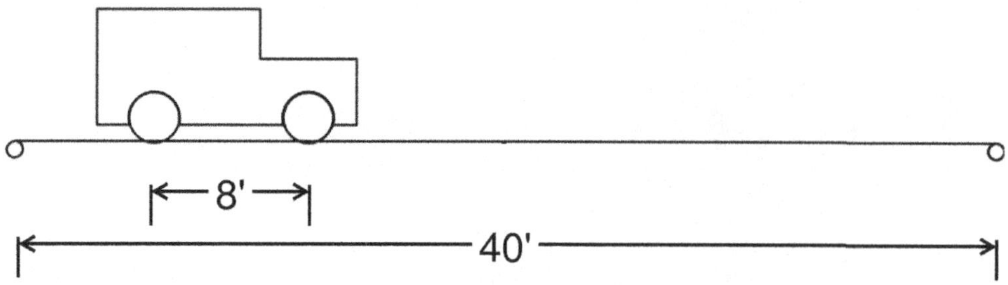

8'
40'

(A) 14.1 K
(B) 18.2 K
(C) 20.4 K
(D) 22.0 K

Question S53

A non-prestressed concrete slab is being formed into a 3′ x 4′ interior column. The thickness of the slab is 14″ and the depth of reinforcing is 12″. Determine most nearly the two-way concrete shear strength of the slab using an f'_c = 4000 psi for normal weight concrete.

Design Code: ACI 318- *Building Code Requirements for Structural Concrete 2014*

(A) 492 K
(B) 550 K
(C) 660 K
(D) 720 K

Question S54

An analysis of an interior girder of a curved bridge determines the following torsional moments: M_{DC1} = 25 K-ft, M_{DC2} = -5 K-ft, M_{DW1} = 10 K-ft, M_{DW2} = -2 K-ft and M_{LL} = 50 K-ft. Determine the factored torsional moment for Strength I.

Design Code: AASHTO – *LRFD Bridge Design Specifications, 7th Ed.*

(A) 100 K-ft
(B) 128 K-ft
(C) 175 K-ft
(D) 220 K-ft

Question S55

For a 3' X 4' column, determine the maximum stirrup spacing for #4 stirrups and #8 longitudinal bars.

Design Code: ACI 318- *Building Code Requirements for Structural Concrete 2014*

(A) 16"
(B) 24"
(C) 36"
(D) 48"

Question S56

A W14x145 has end conditions of pinned and pinned with a length of 38'. The member is braced only about the Y-Y axis 20' from one end. Determine the available LRFD strength in axial compression.

Design Code: AISC – *Steel Construction Manual, 14th Ed.*

(A) 734 K
(B) 1310 K
(C) 1470 K
(D) 1550 K

Question S57

An A36 steel plate as shown below is subjected to a tension load. What minimum thickness is required to resist a factored load of 100 kips?

Design Code: AISC – *Steel Construction Manual, 14th Ed.*

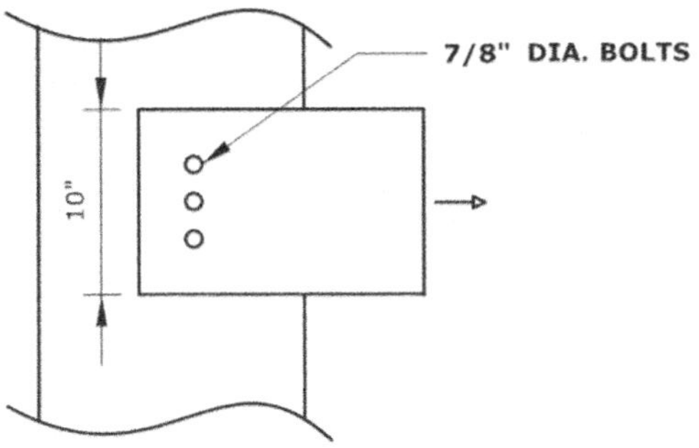
7/8" DIA. BOLTS
10"

(A) 1/4"
(B) 3/8"
(C) 1/2"
(D) 3/4"

Question S58

A simply supported beam has a length of 20' and is to be designed to carry a factored distributed load of 2 K/ft and a factored concentrated load of 10 kips at midspan. Determine which W-shape is most appropriate for the design of flexure about its major axis only for Grade 50 steel. The beam is unbraced full length.

Design Code: AISC – *Steel Construction Manual, 14th Ed.*

(A) W8x48
(B) W10x45
(C) W14x43
(D) W40x167

Question S59

Determine the design stress range for a fatigue category B detail and a design life of 50 years. The detail sees 100 fluctuations per day.

Design Code: AISC – *Steel Construction Manual, 14th Ed.*

(A) 10.25 ksi
(B) 16.0 ksi
(C) 18.67 ksi
(D) 25.55 ksi

Question S60

A 6x6-24F-V8 Glued Laminated Southern Pine Beam has a fully laterally supported length of 20'. The Moisture content is 20% and the beam is designed under normal temperature conditions. Determine the allowable design bending stress. The controlling load case is a uniformly distributed dead plus live plus snow load.

Design Code: NDS – *National Design Specifications for wood construction, 2015 Ed.*

(A) 2120 psi
(B) 2208 psi
(C) 2350 psi
(D) 2400 psi

Question S61

A 6" x 30" southern Pine wood beam is subject to a distributed load of 0.5 k/ft as shown below. Determine the design shear force for the beam.

Design Code: NDS – *National Design Specifications for wood construction, 2015 Ed.*

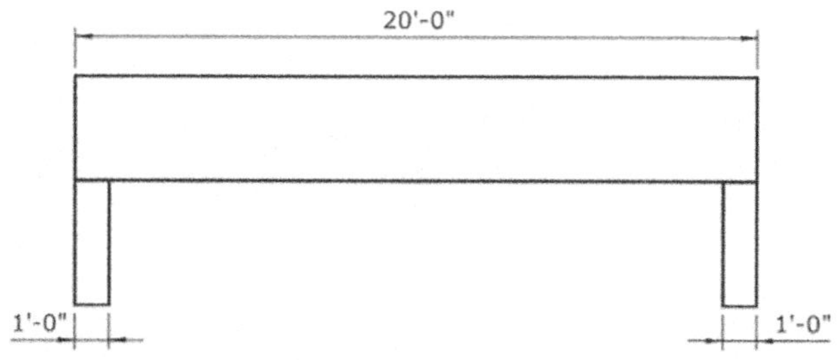
20'-0"
1'-0"
1'-0"

(A) 3.25 K
(B) 5.0 K
(C) 10.5 K
(D) 20.0 K

Question S62

A grouted concrete block masonry beam has a width of 6.5" and a depth of reinforcing of 36". Determine most nearly the service moment capacity using Grade 60 reinforcement with an area of 1.0 in^2. The masonry compressive strength is 1500 lb/in^2.

Design Code: ACI 530 – *Building Code Requirements and Specifications for Masonry Structures, 2013*

(A) 72.67 ft - K
(B) 89.5 ft - K
(C) 95.56 ft - K
(D) 104.04 ft - K

Question S63

Determine the design axial strength of a 14" X 14" reinforced concrete masonry column with a height of 20'. Use Grade 60 reinforcement with an area of steel of 1.76 in^2 and a compressive strength of 2000 psi.

Design Code: ACI 530 – *Building Code Requirements and Specifications for Masonry Structures, 2013*

(A) 109 K
(B) 155 K
(C) 205 K
(D) 250 K

Question S64

Which of the following is not a requirement for shear reinforcement in a concrete masonry beam?

Design Code: ACI 530 – *Building Code Requirements and Specifications for Masonry Structures, 2013*

(A) Shear reinforcement must be parallel to the direction of the applied shear force
(B) Shear reinforcement shall extend a distance of d from the extreme compression face
(C) Shear reinforcement must create a closed loop
(D) The spacing for shear reinforcement must not exceed 1/2 of the beam depth nor 48"

Question S65

Which of the following is not a reinforcement requirement for the design of reinforced masonry?

Design Code: ACI 530 – *Building Code Requirements and Specifications for Masonry Structures, 2013*

(A) The maximum size of reinforcement shall be No. 11
(B) The strength of reinforcing shall not exceed 65000 psi
(C) The diameter of the reinforcement shall not exceed 1/2 of the least clear dimensions of the cell, bond beam, or collar joint
(D) Clear distance shall not be less than the nominal diameter of bar or 1"

Question S66

For the sign support shown, determine the service moment about the X-X axis at the base of the support. The weight of the 5' X 5' sign is 1 kip. Neglect any wind force on the support itself.

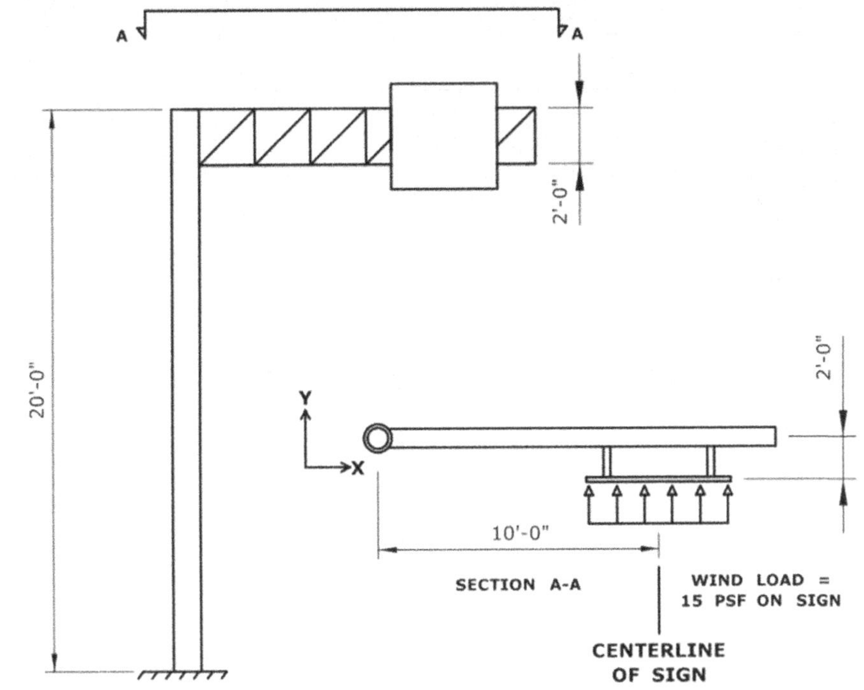
A
A
2'-0"
20'-0"
Y
X
2'-0"
10'-0"
SECTION A-A
WIND LOAD = 15 PSF ON SIGN
CENTERLINE OF SIGN

(A) 5.125 K-ft
(B) 6.0 K-ft
(C) 7.125 K-ft
(D) 11.125 K- ft

Question S67

Determine the Axial force in Member B-D in the truss shown below. Neglect the self-weight of the truss. The support conditions are pinned at A and a roller at B.

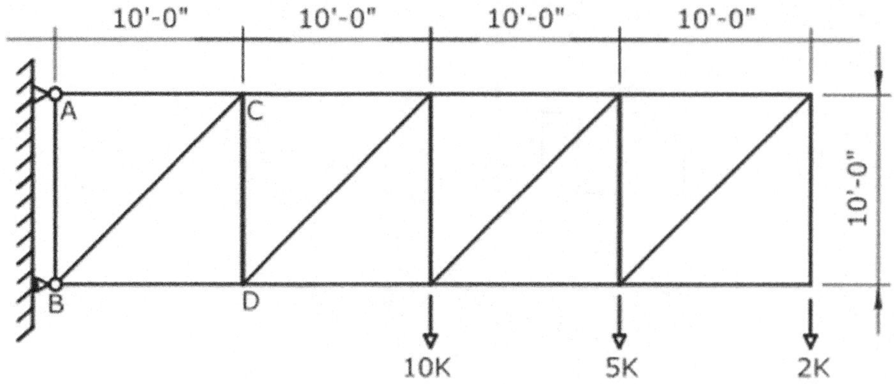

10'-0"
10'-0"
10'-0"
10'-0"
A
C
B
D
10'-0"
10K
5K
2K

(A) 17 K
(B) 24 K
(C) 26 K
(D) 43 K

Question S68

Which of the following would be considered as a DW dead load?

Design Code: AASHTO – *LRFD Bridge Design Specifications, 7th Ed.*

(A) Asphalt driving surface
(B) Beam Self Weight
(C) 6" Water Main
(D) Concrete Deck

Question S69

A 3-span continuous beam is loaded uniformly on the first and last span only. Determine most nearly the maximum moment for a load of 2 k/ft and a span length of 20'.

(A) 80.8 ft-K
(B) 150.0 ft-K
(C) 250.50 ft-K
(D) 970.0 ft-K

Question S70

A Beam with a hinge at B is shown below. Determine the vertical reaction at C.

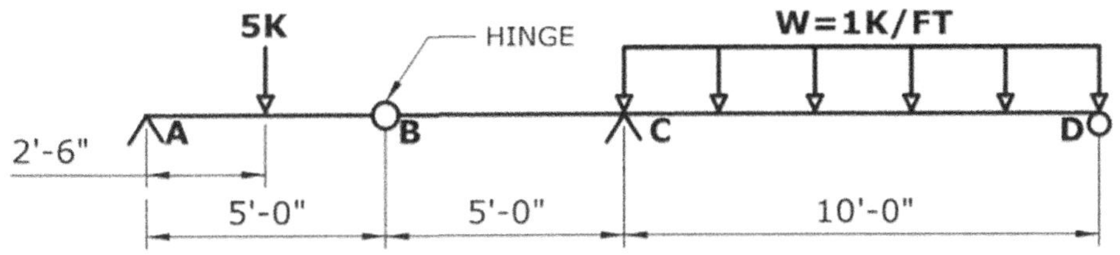

5K
HINGE
W=1K/FT
2'-6"
A
B
C
D
5'-0"
5'-0"
10'-0"

(A) 2.5 K
(B) 4.5 K
(C) 6.0 K
(D) 8.75 K

Question S71

Steel H-Piles are being used to support the pile cap as shown below. Determine the service compressive force in pile A.

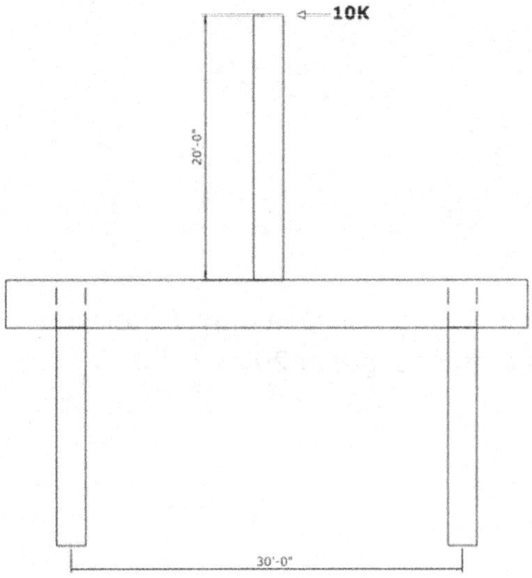

10K
20'-0"
30'-0"

(A) 6.67 K
(B) 10.0 K
(C) 20.0 K
(D) 200.0 K

Question S72

A force of 5 Kips is applied to a bent steel pipe. The plan and elevation views are as shown below. Determine most nearly in Kip-FT the moments about the X, Y, and Z axis. The angle theta can be taken as 32 degrees.

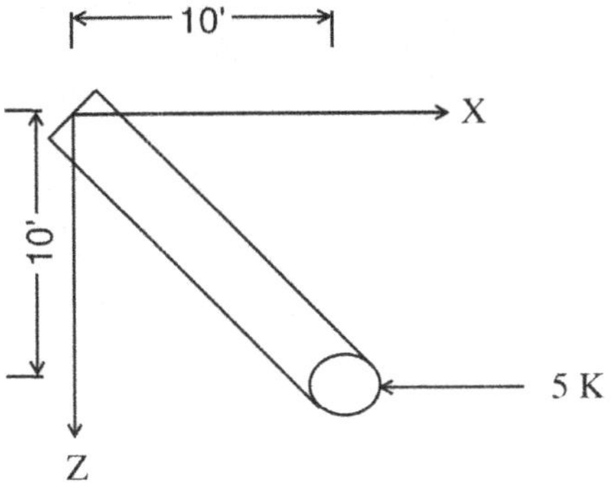
10'
X
10'
Z
5 K

PLAN

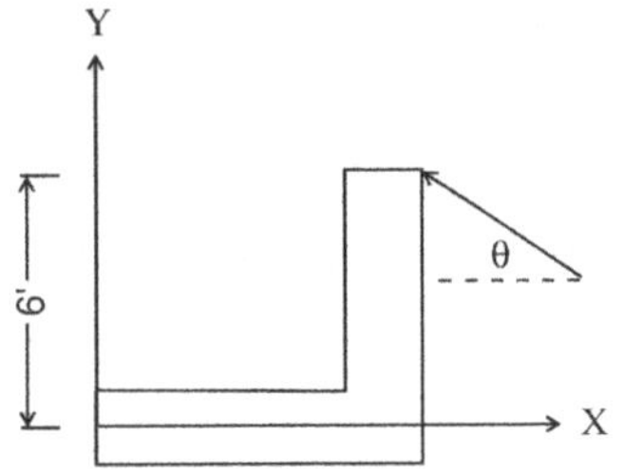
Y
6'
θ
X

ELEVATION

(A) 26.5, 26.5, 51.94
(B) 50, 50, 30
(C) 26.5, 42.4, 51.94
(D) 50, 50, 80

Question S73

Identify all of the following verifications and inspections for concrete structures that are required on a continuous basis?

Design Code: IBC – *International Building Code, 2015 Ed.*

(A) Erection of precast members inspection
(B) Slump and air content tests
(C) Inspection of reinforcing
(D) Inspection of reinforcing steel weldability
(E) Shotcrete placement for proper application techniques
(F) Inspect prestressed concrete for application of forces
(G) Inspect formwork shape

Question S74

A spread footing is to be designed to resist a point load of 40 kips and will have a width of 6'. If the allowable bearing pressure is 2 ksf, determine the minimum length of footing if the distance from the top of the soil to the top of footing is 2' and the thickness of the footing is 3'. Use a soil density of 75 pcf and a concrete density of 150 pcf.

(A) 4.8'
(B) 5.5'
(C) 6.0'
(D) 8.0'

Question S75

A structural cable is used to support 2 point loads as shown below. Determine the tension in section BC of the cable.

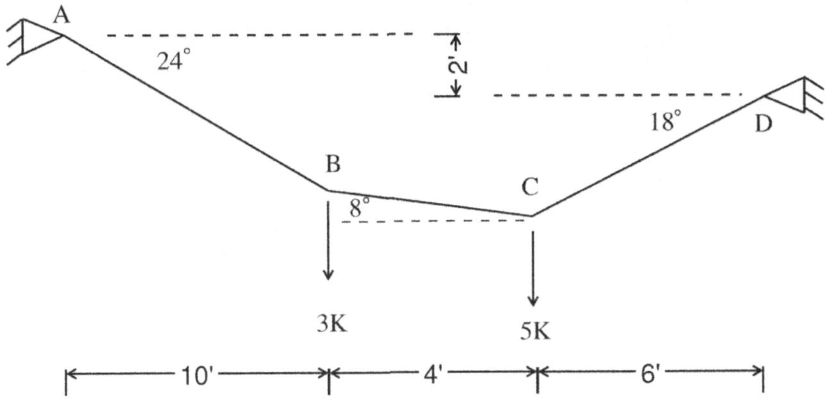
A
24°
2'
18°
D
B
C
8°
3K
5K
10'
4'
6'

(A) 8.5 K
(B) 11.9 K
(C) 12.8 K
(D) 22.2 K

Question S76

A rectangular concrete beam is loaded as shown in the figure below. Determine the bending stress along line A-B as shown in the cross section. Neglect the self-weight of the beam.

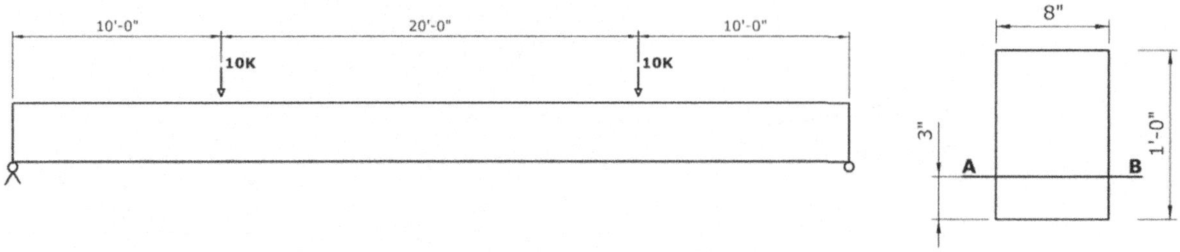

10'-0"
20'-0"
10'-0"
10K
10K
8"
3"
A
B
1'-0"

(A) 2.25 ksi
(B) 3.125 ksi
(C) 4.50 ksi
(D) 5.25 ksi

Question S77

Which of the following loading configurations are included in HL-93?

Design Code: AASHTO – *LRFD Bridge Design Specifications, 7th Ed.*

(A) H-20
(B) HS-20
(C) HS-20 + distributed load
(D) H-20 + distributed load

Question S78

A 100′ column is fixed at the base and free to rotate but not translate at the top. Lateral braces are provided at 50′ and 75′ from the base. Determine the design effective length of the column in the plane of the bracing.

(A) 25′
(B) 40′
(C) 50′
(D) 70′

Question S79

Find most nearly the deflection at the tip of the stepped concrete beam shown below. Neglect the self-weight of the beam and use f'_c = 4000 psi.

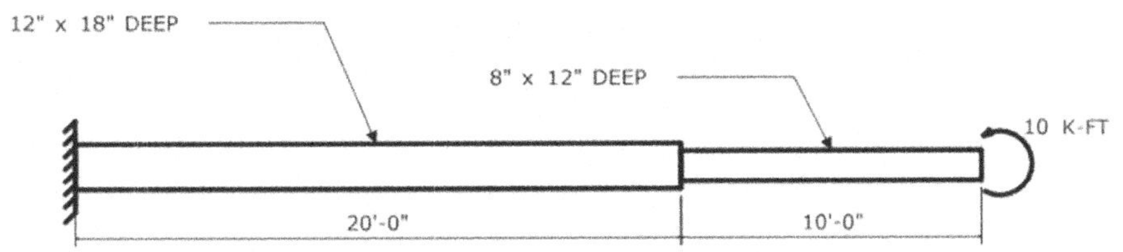
12" x 18" DEEP
8" x 12" DEEP
10 K-FT
20'-0"
10'-0"

(A) 0.25″
(B) 0.33″
(C) 0.54″
(D) 1″

Question S80

A 50′ beam is fixed at one end and is subjected to a tension load of 100 K. Determine the elongation for a W14x109 due to the load.

(A) 0.031″
(B) 0.045″
(C) 0.0060″
(D) 0.065″

SOLUTIONS

Solution M1

Use the average end area method

$V = L(A_1+A_2)/2$ where the length between each station is 50'

1+00 to 1+50:

$V= 50(0+155)/2=3875\ ft^3$

1+50 to 2+00:

$V= 50(155+170)/2=8125\ ft^3$

2+00 to 2+50:

$V= 50(170+65)/2=5875\ ft^3$

2+50 to 3+00:

$V= 50(65+0)/2= 1625\ ft^3$

Then simply add the volumes and convert to cubic yards

$3875+8125+5875+1625 = 19500\ ft^3\ (0.037037\ yd^3/ft^3) = 722.22\ yd^3$. The Answer is **(A)**

Solution M2

A basic knowledge of formwork is needed for this problem. Since the footing has already been poured, formwork will only be needed for all of the faces of the wall except for the top of the wall. The solution is to add up the surface area of the wall faces.

Rear face walls

$13.5'(8.0') + 10.0'(8.0') = 188\ ft^2$

Front face walls

$(10.0' - 1.5')8.0' + (13.5'-1.5')8.0' = 164\ ft^2$

Outer walls

$(1.5'(8.0'))(2\ walls) = 24\ ft^2$

Add up the surface areas = $188 + 164 + 24 = 376\ ft^2$. The answer is **(C)**

Solution M3

The Critical path of a schedule is the sequence of tasks that determine the minimum amount of time needed to complete the project. Any task on a critical path whose duration is changed will affect the overall schedule of the project.

To solve first develop the task flow chart as shown below. To do this go to each task individually and ask the question "which tasks have this one as a predecessor?". Then draw and connect those tasks with arrows to create a path. Do the same with each task individually until you reach the final task. Then add up the durations for each potential path and the critical path is the largest duration.

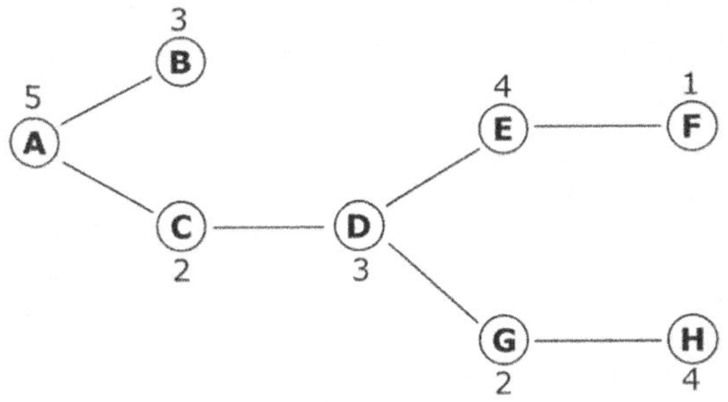

3
B
5
A
4
E
1
F
C
2
D
3
G
2
H
4

The critical path is tasks A-C-D-G-H with a duration of 16 months.

Solution M4

The initial cost is already at present worth so there is no adjustment of value. The maintenance costs each provide a future value which needs to be converted into a present value

What we have:

n=10 and 20
F= 10,000 and 15,000
i= 3%

What we need: P

Equation needed: $P = F(1+i)^{-n}$

1st maintenance cost: $10{,}000(1+0.03)^{-10} = 7441$

2nd maintenance cost: $15{,}000(1+0.03)^{-20}= 8305$ Then add together all present worth costs = 100,000 + 7441 + 8305 = 115746. The answer is **(D)**

Solution M5

First determine the plasticity index: PI = LL – PL = 55-23 = 32

The 50% finer than No. 4 sieve indicates a coarse-grained sandy soil.

Using the word "fines" in a problem is another way to define the No. 200 sieve. Since 20% is greater than 12%, this narrows it down to SM or SC.

A PI over 7 then indicates SC. The Answer is **(A)**

Solution M6

The maximum moment of a simply supported beam is at midspan. Add together the moment from the distributed load and the point load. This can be done by statics or by using the predetermined equations. It's always best to save time by using design aids.

Moment from distributed load = $wL^2/8 = 2(20^2)/8=100$ k-ft
Moment from point load = $PL/4 = 3(20)/4 = 15$ k-ft
Total moment = 100 + 15 = 115 k-ft. The Answer is **(B)**

Solution M7

The end conditions dictate the effective length of a compression member. From left to right the correct theoretical factors are 2.0, 0.5, 0.7, and 1.0.

Solution M8

The first step is to determine the magnitude and direction of the support reactions

Sum moments about the first support A = 5(5') + 10(10') + 5(15') – B(20') Solve for the right support, B:

B=10 Kips, A = (5 + 10 + 5) – 10 = 10 kips

Then you are able to layout the shear diagram using the following rules:

- To find the magnitude at any point, take a free body diagram from that point to the left-most support and add up the reactions.
- Reactions and loads pointing up are positive, those pointing down are negative
- The shear diagram is flat between concentrated loads
- The shear diagram is sloping along distributed loads

Begin constructing at the left of the beam and work right. The first point is equal to the reaction at A and is upward so the first point is at 10 kips. There is no change until at 5' along the length there is a loading of 5 kips. Therefore, the graph drops to 10 – 5 = 5. Follow this trend to develop the graph.

The answer is **(A)**

Solution M9

First determine the reactions at the left support, A:

Sum moments about point the right support, B = A(10') – 2 k/ft(10')(5') - 5'(2), A = 11 Kips

Since there is no easy equation for the moment at this point on this beam, use the following rule about moment diagrams:

- The magnitude of the moment at any point is equal to the area under the shear diagram curve up to this point

Begin by constructing the shear diagram. We can save time by stopping at point C since we don't need to know any additional information. The graph begins with the reaction at A of 11 kips. You can then find the shear at point C by subtracting the magnitude of the load up to this point.

Shear at C = 11 – 2 k/ft(4') = 3 kips. Therefore, the graph now looks like the figure below. Divide it into sections and find the area:

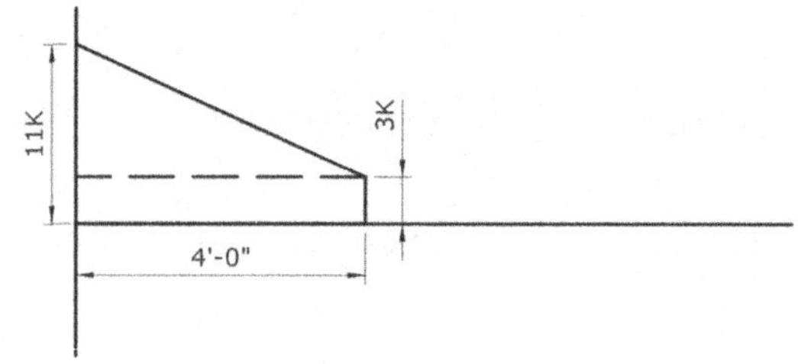
11K
3K
4'-0"

Area 1 = 3 k(4') = 12 kip-ft
Area 2 = ½ (11k - 3k)(4') = 16 kip-ft Total Moment = 12 + 16 = 28 K-ft The answer is **(C)**

Solution M10

6" diameter concrete cylinders are a common method of testing concrete compressive strengths. The stress from an axial load is determined by P/A.

Area of each cylinder = $\pi r^2 = \pi(3)^2 = 28.27$ in^2

Cylinder 1 stress = 102000/28.27 = 3607 psi
Cylinder 2 stress = 111000/28.27 = 3925 psi
Cylinder 3 stress = 100500/28.27 = 3554 psi

Average of the cylinders = (3607 + 3925 + 3554)/3 = 3695 which is most nearly 3700 psi.

The answer is **(D)**

Solution M11

You can find the length of the curve by subtracting stations L = 2+25 – 1+50 = 75'
Convert the interior angle to a decimal 15'/60 = 0.25 Therefore I = 10.25°
The radius can be found by $L = (2\pi rI)/360° = 75' = 2\pi r(10.25)/360$ solve for r = 419.2'

The answer is **(B)**

Solution M12

The clearance of the bridge is simply the elevation difference between the low point of the bridge and the elevation of the roadway at that station.

First calculate the gradient of the curve $A = (G_2 - G_1)/L = (-4-2)/8 = -0.75$

Determine the beginning of vertical curve (BVC) station:

BVC = 10+00 – 800/2 = 6+00

Determine the elevation of BVC = 155.0 – (.02)400 = 147.0'

Determine the elevation of station 8+00 by using the equation of a parabola (Be careful with signs) where X is the difference in stations from the point of interest to the BVC

$Y = Elev_{BVC} + G_1(X) + (A/2)X^2 = 147.0 + 2(2) + (-0.75/2)(2)^2 = 149.5'$

Then subtract the road elevation from the low point of the bridge to find the clearance

170.0 – 149.5 = 20.5'.

The answer is **(B)**

Solution M13

The low point is where the slope of the gradient is zero. Therefore:

$X = G_1/A$

$A = (G_2-G_1)/L = (2-(-3))/6 = 0.833$

X = 3/0.833 = 3.6 stations

The low point is at station 7+00 + 3+60 = 10+60.

Solution M14

Use the equation for the middle ordinate of a horizontal cure

M=R(1-cos(28.65S/R)) = 700(1-cos(28.65(220)/700)) = 8.63'

The answer is **(A)**

Solution M15

Working backwards from compaction, apply adjustment factors at each stage and carry over the volume:

To achieve 100 yards of 90% compaction you will need $100/0.9 = 111.11\ yd^3$

5% is lost so you will need an additional 5% to account for this $111.11(1.05) = 116.65\ yd^3$

A swell factor is the volume of the loose excavation material to the in-place excavation material
To find the in-place material, divide by the swell factor $116.65/1.07 = 109.03\ yd^3$

The answer is **(D)**

Solution M16

To solve mix design problems, follow the units.

Determine the weight needed in water $W_{water} = W_{cement}(w/c\ ratio) = 500(0.65) = 325\ lb/yd^3$

The specific weight of water can always be taken as $62.4\ lb/ft^3$. Use the specific weight to convert a weight to a volume

$V_{water} = W_{water}/\gamma_{water} = (325\ lb/yd^3)/(62.4\ lb/ft^3) = 5.21\ ft^3/yd^3$

The answer is **(A)**

Solution M17

There are 3 conditions which will simplify the needed equations.

1. Friction is neglected between the wall and soil. This make the angle of external friction $\partial=0$
2. The backfill is horizontal. This makes the slope of backfill $\beta=0$
3. The wall face is vertical. This makes $\theta=0$

The equations needed are now

The active earth pressure coefficient $k_a=\tan^2(45^\circ-\phi/2)$
The total active resultant $R_a= ½\ k_a\gamma H^2$

$k_a=\tan^2(45-25/2)=0.406$
$R_a= ½\ (0.406)(115pcf)(15.5)^2= 5606$ lbs

The force is applied at H/3=15.5/3 = 5.167'

The moment is 5.606k(5.167') = 28.9 k-ft

The answer is **(C)**

Solution M18

The void ratio is the volume of the voids divided by the volume of the solids.

If a sample is saturated it can be assumed that there are no air voids and thus the total volume of the voids is equal to the volume of the water in that state.

When a sample is dried, there is no more water left in the sample and we have simply the weight of the soil.

Therefore, we are left with the following values

$W_W = 50 – 42 = 8$ lbs

$W_S = 42$ lbs

Since the density of water is known we can then convert the weight to volume

$V_W = 8/62.4 = 0.128$

And since the sample was saturated, all voids are filled with water and $V_W=V_V$

We then need to determine the volume of soil from the weight. Since we have the Specific Gravity (SG), we can determine the density by multiplying by the density of water

$\gamma = 2.4(62.4) = 150$ lb/ft^3

$V_S = 42/150 = 0.28$

Void Ratio = 0.128/0.28 = 0.46

The answer is **(D)**

Solution M19

The total distance is a sum of two components. The first is before breaking and the second is after breaking. This is represented in the following equation:

$S_{stopping}=vt_p+s_b$

The first component assumes the velocity is constant during perception reaction and is simply (100 ft/s)(2 seconds) = 200 ft

The stopping distance when breaking occurs is the following:

$s_b = v^2_{mph}/(30(f+G))$

Since the slope is on an incline G is positive. If it was on a decline, it would be negative.

Convert the velocity from ft/s to mi/hour

100ft/sec(3600 sec/hr)/5280 ft/mi = 68.2 mi/hr

$s_b = (68.2^2)/(30(0.30+0.02)) = 484.5$ ft

$S_{stopping} = 200 + 484.5 = 684.5$ ft

The answer is **(B)**

Solution M20

The superelevation transition distance is the distance between the beginning of the transition to superelevation to the point of being fully superelevated which is the combination of the tangent (also known as crown) runout and the superelevation runoff.

The answer is **(A)**

Solution M21

Use the Darcy equation:

$h_f = (fLv^2)/(2Dg)$

Use 32.2 ft/sec^2 as g and then plug in and solve for v

$3.0 = (0.02)(500)v^2/2(2)(32.2)$

v = 6.2 ft/s The answer is **(C)**

Solution M22

The appropriate equation in open channel flow is the Chezy-Manning equation:

$Q = (1.49/n)AR^{2/3}S^{1/2}$

First calculate the hydraulic radius which is the area of water divided by the wetted perimeter which is the perimeter of the sides of the channel which are in contact with water.

$R = (6*3)/(6+3+3) = 1.5$

$Q = (1.49/0.015)(6*3)(1.5)^{2/3}0.003^{1/2} = 129.3$ The answer is **(A)**

Solution M23

For the given information, the Hazen-Williams equation is appropriate:

$h_f = 10.44LQ^{1.85}/C^{1.85}d^{4.87}$

$40=10.44(100)Q^{1.85}/(140)^{1.85}(3)^{4.87}$

Solve for Q = 432 gpm The answer is **(D)**

Solution M24

The time base is the amount of time that the flow exceeds the base flow. The answer is **(C)**

Solution M25

The flow rate of the pipe can be used to determine the area required to limit the flow velocity

The flow rate of the pipe can be determined by the conservation of flow principle

$Q_1 + Q_2 = Q_3$

The flow of the drainage areas can then be determined by the Rational method:

$Q = ACi$ Therefore $Q_3 = 15(0.18)(1.5) + (10)(0.22)(1.5) = 7.35$ cfs

Use the calculated velocity to find the required area $Q/V = A = 7.35/0.26 = 28.28\ ft^2$

The answer is **(A)**

Solution M26

When determining how many 0-Force members a truss has, analyze each joint individually as a free body diagram and follow these guidelines:

- In a joint with 2 members and no external forces or supports, both members are 0-force
- In a joint with 2 members and external forces, If the force is parallel to one member and perpendicular to the other, then the member perpendicular to the force is a 0-force member.
- In a joint with 3 members and no external forces, if 2 members are parallel then the other is a 0-force member

Therefore, each joint can be analyzed as follows:

Joint A and H –These joints have 2 members and an external force so refer to guideline 2. Neither member parallel to the force therefore all members are non-zero

Joint C, D, and G – These joints have 3 members and no external force so refer to Guideline 3. 2 members are parallel therefore the other is a 0-Force member. Therefore, Members C-B, D-E, and F-G are 0-Force members

Joints B, E, and F – These joints do not meet any of the guidelines. Therefore, all members are non-zero

Solution M27

This question is provided simply to illustrate that the analysis of the external forces of a truss can be determined in the same way as beams. There may be a tendency to perform unnecessary computations. Simply apply basic statics:

Sum the Forces about H = 0 = -40A + 5(30) + 5(10), Solve for A = 5 Kips

The Answer is **(B)**

Solution M28

First determine the reaction at A. See solution M27.

Determine the angle of triangle A-B-C $\text{Tan}^{-1}(8/10) = 38.66$

Using the method of joints, analyze joint A using a free body diagram

Sum the forces in the Y-Direction 5 = ABsin(38.66), AB = 8.00

Sum the forces in the X-Direction AC = ABcos(38.66) = 8cos(38.66) = 6.24

Analyze Joint C

Sum the Forces in the X-Direction AC = CE = 6.24

The answer is **(C)**

Solution M29

Rebar corrosion is most often the result of a chemical reaction due to the intrusion of chlorides into existing concrete. The answer is **(A)**

Solution M30

The water to cement ratio is inversely proportional to the concrete strength and therefore has a direct impact. The answer is **(A)**

Solution M31

Use the Bernoulli equation for the conservation of energy

$E_t = E_{pr} + E_v + E_p = p + v^2/2g + z$

At the water surface of the reservoir, the total energy is the potential energy only and $E_t = 0 + 0 + 50 = 50$

Due to the conservation of energy the total energy at the water surface is equal to the total energy at the exit of the pipe. Therefore, the total energy at the exit
$E_t = 50 = 0 + v^2/2(32.2) + 0$, solve for v = 56.75 ft/s The answer is **(B)**

Solution M32

The answer is **(D).**

The other 3 assumptions are essential to the Bernoulli equation.

Solution M33

Use the conservation of flow principle. $Q_1 + Q_2 = Q_3$ and therefore $A_1V_1 + A_2V_2 = Q_3$

$Q_3 = 3.0(1.2) + (4.0)0.8 = 6.8$ ft^3/s The answer is **(D)**

Solution M34

The effective stress is the density times the height of each level. However, if the water table is present, the density is reduced by that of water:

Effective Stress = 130(20) + 100(10) + (100-62.4)(10) = 2600 + 1000 + 376 = 3976 psf

The answer is **(C)**

Solution M35

The Maximum bending stress is determined at the location of maximum moment.

First determine the reactions at the supports

Sum moments about B = 0 = -15A + 10(10) + 10(5), A = 10

Due to symmetry A = B = 10

The Max moment for a beam with 2 symmetrical point loads = PX where X is the distance to the first point load = 5'(10 kips) = 50 Kips-ft

Then apply the equation for bending stress Mc/I

Moment of inertia of a rectangle $(1/12)bh^3 = (1/12)(4)(6)^3 = 72\ in^4$

The bending stress = $(50\text{ k-ft})(12\text{ in/ft})(3\text{ in})/72\ in^4 = 25$ ksi

The answer is **(A)**

Solution M36

P/A = 0.1 ksi = 9/(12w), w = 7.5"
B can be determined by finding the additional horizontal distance from the critical section for shear to the base.

$$\frac{1}{8} = \frac{x}{6} \quad x = 0.75"$$

b = 7.5" + 0.75" = 8.25"

The answer is **(C)**

Solution M37

The determination of which roadside safety barrier system is determined by the distance from the edge of road. Barriers can be flexible, semi rigid, or rigid. A concrete barrier is rigid and is for locations where zero deflection is needed. A metal beam rail is semi-rigid and is used for objects or drop-offs within a short distance of the road. Cable systems are flexible and have the most deflection and are used for objects or drop-offs further from the road.

The answer is **(C)**

Solution M38

The peak hourly traffic volume is the hour-long timeframe in which the most cars are observed. First calculate the total volume for each hour-long time frame:

8:00-9:00 = 500 + 560 + 650 + 625 = 2335
8:15-9:15 = 560 + 650 + 625 + 630 = 2465
8:30-9:30 = 650 + 625 + 630 + 600 = 2505
8:45-9:45 = 625 + 630 + 600 + 540 = 2395
9:00-10:00 = 630 + 600 + 540 + 460 = 2230

Then determine the largest volume = 2505 The answer is **(C)**

Solution S39

The equation for deflection at the end of a cantilever beam with a point load is:

$Pl^3/3EI$

First determine the moment of inertia:

$I = (1/12)(6)12^3 = 864\ in^3$

$\Delta = 1(10(12))^3/(3(3605)(864)) = 0.18''$, The answer is **(B)**

Solution M40

The specific gravity is the density of the soil over the density of water.

The volume of the voids can be determined by subtracting the soil volume from the total

$V_V = V_T - V_S = 2.0 - 1.5 = 0.5\ ft^3$

The degree of saturation can be used to determine the volume of water

$V_W = V_V(S) = 0.5(0.75) = 0.375\ ft^3$ Convert volume to weight $W_W = 0.375(62.4) = 23.4$ lb

Then determine the weight of soil thought the moisture content

$W_S = W_W/w = 23.4/0.1 = 234$ lbs, the density is then 234/1.5 = 156

The SG = 156/62.4 = 2.5 The answer is **(B)**

Solution S1

It is important to note that the flat roof snow load is different from ground snow load and design roof snow load.

$P_f = P_g(0.7C_eC_tI_S)$

C_e = Exposure Factor from Table 7-2 = 0.9
C_t = Thermal Factor from Table 7-3 = 1.2
I_S = Importance factor from 1-5.2 = 1.0

$P_f = 35(0.7)(0.9)(1.2)(1.0) = 26.5$ psf

Ensure this is greater than the minimum $= 20I_S = 20(1.0) = 20 < 26.5$ psf The answer is **(C)**

Solution S2

First calculate the flat roof snow load

$P_f = P_g(0.7C_eC_tI_S)$

C_e = Exposure Factor from Table 7-2 = 0.9
C_t = Thermal Factor from Table 7-3 = 1.2
I_S = Importance factor from 1-5.2 = 0.8

$P_f = 40(0.7)(0.9)(1.2)(0.8) = 24.2$ psf

The design roof snow load $P_S = C_SP_f$
Determine C_S from Figure 7-2c. This is a cold roof with a non-slippery surface due to the asphalt shingles. Therefore $C_S = 1.0$. The answer is **(B)**

Solution S3

The seismic base shear $V = C_sW$

C_S = Seismic response coefficient $= S_{DS}/(R/I_e)$

Determine I_e from table 1.5-2 using the risk category = 1.0

R = 3.0 from table 12.2-1

$C_S = 0.3/(3.0/1.0) = 0.1$

V = 0.1(200) = 20 K. The answer is **(B)**

Solution S4

The equation is $S_{DS} = 2/3S_{MS} = 2/3F_aS_S$

S_S is given = 0.5, and F_a can be determined from the site classification.

From Table 20.3-1 the site classification is C, Therefore from Table 11.4.1 $F_a = 1.2$

$S_{DS} = 2/3(1.2)(0.5) = 0.4$ The answer is **(C)**

Solution S5

The multiple presence factor is used to account for the probability of the number of design lanes being loaded at the same time. It is only a function of the number of lanes loaded for the analysis. As per table 3.6.1.1.2-1, the answer is **(A)**

Solution S6

ASCE Table 12.6-1 indicates permitted analytical procedures for different building types. Option D includes a building in Category D exceeding 160' and $T > 3.5T_S$. Therefore this falls under "All other structures" and equivalent lateral force procedure is not permitted. The answer is **(D)**

Solution S7

From Table 11.6.1 the Seismic Category can be determined. The category is B

Solution S8

The dead load is given as 30 psf
The Live Load is determined given the occupancy of a hospital used for operating LL = 60 psf
Use the load combinations to determine the factored design load. By inspection use 1.2DL + 1.6LL

1.2(30) + 1.6(60) = 132 psf The answer is **(A)**

Solution S9

From the occupancy, determine the Live Load = 40 psf. Determine the Live Load reduction where:

$L = L_O(0.25 + 15/\sqrt{K_{LL}A_T}\ )$ $A_TK_{LL} = 4(300) = 1200 > 400$ OK

$L = 40(0.25 + 15/\sqrt{1200}) = 0.683(40) = 27$ psf $> 0.4L_O$. OK

The answer is **(A)**

Solution S10

The reaction at the left support A is determined by taking moments about the right support B. However, the maximum effect includes the load combination which would yield the largest force effect. The unfactored force effects are as follows:

D = (5+10)10 = 150
L = 15(10) = 150
L_R = 2(10) = 20
S = 4(10) = 40
E = 8(30) = 240
W = 2(30) = 60

Then determine which load combination is the controlling case by plugging in the force effects. By inspection Load cases 1, 6, and 7 are eliminated

Case 2 = 1.2D + 1.6L + 0.5(L_R or S or R) = 1.2(150) + 1.6(150) + 0.5(40) = 440
Case 3 = 1.2D + 1.6(L_R or S or R) + (L or 0.5W) = 1.2(150) + 1.6(40) + 150 = 394
Case 4 = 1.2D + 1.0W + L + 0.5(L_R or S or R) = 1.2(150) + 60 + 150 + 0.5(40) = 410
Case 5 = 1.2D + 1.0E + L + 0.2S = 1.2(150) + 240 + 150 + 0.2(40) = 578

Therefore Case 5 is the controlling case. To find the reaction divide the force effects by the distance between supports as in summing moments about B, A = 578/20 = 28.9 k. The Answer is **(D)**

Solution S11

AWS Table 4.4 provides requirements for WPS (Welding Procedure Specification) Qualifications including testing. The test type not included in the fillet weld type is Reduced Section Tension. The answer is **(D)**

Solution S12

IBC section 1810.2.1 has requirements for the length of deep foundation elements which can be considered laterally supported. For a stiff soil consider the support beginning at 5' from the top of embedment. Therefore 40'-5' = 35' The answer is **(A)**

Solution S13

IBC Table 601 indicates the fire-resistance rating requirements for certain building types. The table is further divided into class A and B. A indicates External protection and B is unprotected. Therefore, the answer is **(C)**

Solution S14

First determine the applicable cross section from table 4.6.2.2.1-1, a steel girder with concrete deck is cross section (a).

Then determine the appropriate equation from table 4.6.2.2.2b-1 considering moment for interior beams. The equation is:

$0.06 + (S/14)^{0.4}(S/L)^{0.3}(K_g/(12Lt_s^3))^{0.1} = 0.06 + (6/14)^{0.4}(6/60)^{0.3}(800600/(12(60)(8)^3))^{0.1} = 0.45$

Note the appropriate units for the equation.

The answer is **(B)**

Solution S15

The design axial strength of short compression members does not incorporate any P-Delta effects and therefore is simply the strength in direct compression of the concrete and steel combined. The equation is:

$\phi P_n = \phi 0.80(0.85f'_c(A_g - A_S) + f_y A_S)$
$A_g = 14(12) = 168\ in^2$, $A_S = 1.0(6) = 6\ in^2$, $\phi = 0.65$ for tied columns
$\phi P_n = 0.65(0.80)(0.85(4)(168\text{-}6) + 60(6)) = 473.6$ K The answer is **(B)**

Solution S16

The development length of a standard 90-degree hook is the following equation as per section 12.5:

$L_{dh} = (0.02\Psi_e f_y/\lambda\sqrt{f'_c})d_b$

$\Psi_e = 1.0$ for non-epoxy coated reinforcement
$\lambda = 1.0$ for normal weight concrete

$L_{dh} = (0.02(1.0)60000/1.0\sqrt{4000})(6/8) = 14.23''$

Now apply appropriate adjustment factors.

Since the cover is sufficient, 0.7 may be applied

Since the area of steel provided = area of steel required, there is no adjustment

14.23(0.7) = 9.96, use 10"

However, the question asks for the length L. It is important to remember the critical section past which the rebar needs to be developed. In this case it is at the face of the column. Therefore
L = 10 + (12 – 2) = 20" The answer is **(D)**

Solution S17

Use Table 9.3.1.1 to determine the appropriate equation. In this case use l/16 = 50(12)/16 = 37.5" However the depth must be modified for reinforcing strengths other than Grade 60.

$0.4 + f_y/100000$ = 0.4 + 40000/100000 = 0.8, Multiply the minimum for Grade 60, 0.8(37.5) = 30" The answer is **(A)**

Solution S18

The effective flange width is the minimum of the following 3 calculations as per table 6.3.2.1:

1. ¼ of the span length = 25(12)/4 + 12" = 87"
2. The width of the beam + 16(thickness of the slab) = 12" + 16(6) = 108"
3. Width of beam + clear distance between beams = 12" + 10'(12) = 132"

The answer is **(A)**

Solution S19

The minimum reinforcing is the larger of the following two equations

$3\sqrt{f'_c}b_wd/f_y = \sqrt{4000}(3)(12)(15)/60000 = 0.57\ in^2$

$200b_wd/f_y = 200(12)15/60000 = 0.6\ in^2$

The maximum reinforcing does not have a simple equation but is a function of limiting the strain in the steel so that the mode of failure is not crushing of the concrete. This is done by setting the strain of steel to 0.005. Therefore:

$0.005 = ((d\text{-}c)/c)(0.003) = ((15\text{-}c)/c)(0.003)$, solve for $c = 5.625$

$a = \beta c$, for $f'c = 4000$, $\beta = 0.85$, $a = 0.85(5.625) = 4.78''$

$a = A_sf_y/(0.85f'_cb) = 4.78 = A_s(60)/(0.85(4)(12))$ solve for $A_s = 3.25\ in^2$

The answer is **(A)**

Solution S20

ACI provides equations in section 6.5.2 for moments in a frame using approximate analysis:

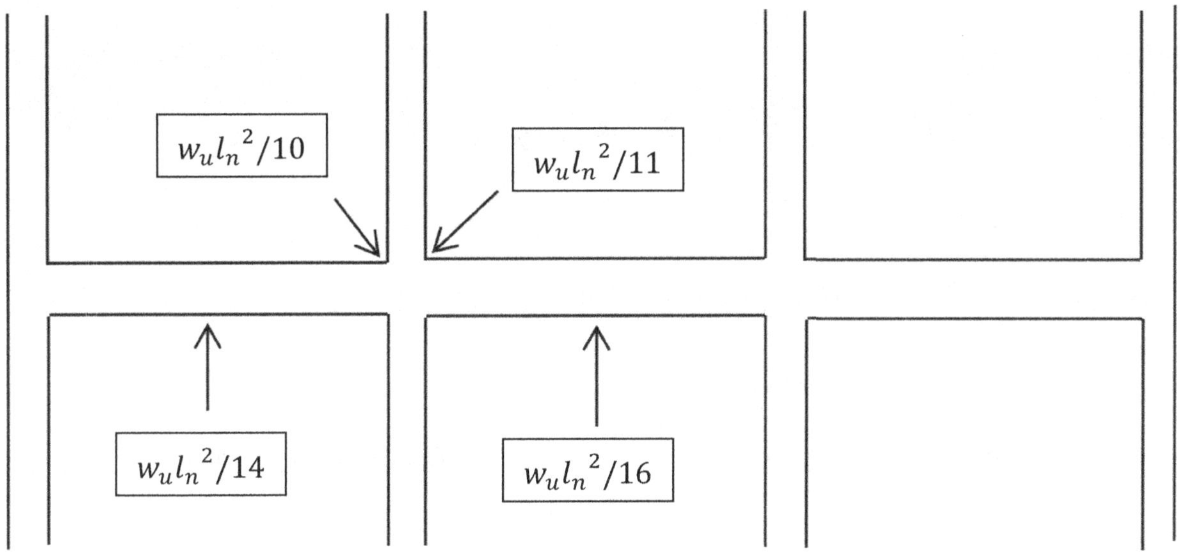

$w_u l_n^2/10$
$w_u l_n^2/11$
$w_u l_n^2/14$
$w_u l_n^2/16$

Solution S21

The types of Prestress losses are the following:

(A) Elastic shortening of concrete
(B) Shrinkage of concrete
(C) Creep of concrete
(E) Friction
(F) Relaxation of steel

Solution S22

The shear capacity of a concrete beam is the addition of the shear strength of the concrete and the reinforcing stirrups. Therefore:

$\phi V_n = \phi V_c + \phi V_s$

$\phi V_c = 2\phi\sqrt{f'_c} b_w d = 2(0.75)\sqrt{4000}(12)(24)/1000 = 27.3$ K

$V_S = (V_u - \phi V_c)/\phi$

Therefore the remaining capacity needed is (70 – 27.3)/0.75 = 56.9 K

$V_s = 62.9 = (A_v f_y d)/s$, s = (2(0.2)(60)(24))/56.9 = 10.1"

Ensure this is less than max spacing = $A_v f_y / 50 b_w$ = 0.2(2)(60000)/50(12) = 34" OK

Check that $V_s < 4\sqrt{f'_c} b_w d$ = 72.8 > 56.9 OK

The answer is **(A)**

Solution S23

The plastic moment is simply the yield strength times the plastic section modulus: $M_P = F_y Z$
The plastic section modulus is the summation of the areas in compression and tension multiplied by the distance from the center of gravity of each area to the plastic neutral axis. Since the member is doubly symmetric, the PNA is about the center. Therefore:

$Z = 2(A_f Y_f + A_w/2 Y_w) = 2(12(0.5)(10+0.5/2) + 20(0.5)/2(20/4)) = 173$ in^3

M_P = 50(173) = 8650 in-k/12 = 720.8 ft-k The answer is **(A)**

Solution S24

$\phi R_n = \phi F_n A_b n$

$A_b = \pi(0.875)^2/4 = 0.601$ in^2

F_n = 68 ksi from table J3.2 with threads excluded

Also, given the connection type, the capacity is doubled due to 2 shear planes

200 = 2(0.75)(68)(0.601)n, n = 3.26, use 4 bolts. The answer is **(B)**

Solution S25

The main concept to understand for block shear is that a component of the capacity comes from tension which is the length of perimeter perpendicular to the load and the other comes in shear from the portion which is parallel to the load. Then simply fill in the equation:

$R_n = 0.6F_uA_{nv} + U_{bs}F_uA_{nt} < 0.6F_yA_{gv} + U_{bs}F_uA_{nt}$

In most cases U_{bs} = 1.0 if tension is uniform (this can always be assumed if not specifically stated)

It's important to note what the subscripts stand for T=Tension face, V=Shear face, N=net area, G=gross area. Then calculate the variables

The diameter of the hole is 1/8" + diameter of the bolt = 1/8 + 7/8 = 1"

$A_{nv} = 2((5+4+4)-2.5(1))0.5 = 10.5$
$A_{nt} = (6-1(1))0.5 = 2.5$
$A_{gv} = 2(13)0.5 = 13$

Then plug in the equation

$R_n = 0.6(58)(10.5) + 1.0(58)(2.5) < 0.6(36)(13) + 1.0(58)(2.5) = 510.4 < 425.8$ Therefore use 425.8K

Don't forget phi

0.75(425.8) = 319.35 K The answer is **(C)**

Solution S26

The equation for the strength of the weld is:

$\phi R_n = \phi 0.6F_{EXX}(0.707wL) = (0.75)(0.6)(70)(0.707)(7/8)(8) = 155.9$ K

However this value is only for welds parallel to the force. The strength needs to be modified by the following equation to account for the angle of the weld to the load:

$0.5\sin^{1.5}\theta + 1$, where $\theta = 90$, $1.0 + 0.5\sin^{1.5}(90) = 1.5$. $\phi R_n = 1.5(155.9) = 233.9$ K

The answer is **(D)**

Solution S27

$F'_c = F_cC_DC_MC_tC_P$
C_D = 1.15 as per table 2.3.2
C_M = 1.0 since moisture content < 16%
C_t = 1.0 under normal temperature conditions

From table 5A find the following reference values:

F_c = 1850 psi
$E_{Xmin} = 1.00 \times 10^6$
$E_{ymin} = 0.95 \times 10^6$

$$C_p = \frac{1 + \frac{F_{cE}}{F*_c}}{2c} - \sqrt{\left(\frac{1 + \frac{F_{cE}}{F*_c}}{2c}\right)^2 - \frac{\frac{F_{cE}}{F*_c}}{c}}$$

$$F_{CE} = \frac{0.822E'_{min}}{\left({}^{l_e}/_{d}\right)^2}$$

c = 0.9 for Structural Glued Laminated

$F*_C = F_cC_DC_MC_t = 1850(1.15)(1.0)(1.0) = 2128$ psi

However C_p needs to be evaluated about both the X-axis and Y-axis
$E'_{Xmin} = E_{Xmin}C_MC_t = 1.00 \times 10^6(1.0)(1.0) = 1.00 \times 10^6$
$E'_{ymin} = E_{ymin}C_MC_t = 0.95 \times 10^6(1.0)(1.0) = 0.95 \times 10^6$

$$F_{CEx} = \frac{0.822(1.00\ x\ 10^6)}{\left({}^{20(12)}/_{12}\right)^2} = 2055$$

$$F_{CEy} = \frac{0.822(0.95\ x\ 10^6)}{\left({}^{10(12)}/_{6}\right)^2} = 1952$$

$$C_{px} = \frac{1 + \frac{2055}{2128}}{2(0.9)} - \sqrt{\left(\frac{1 + \frac{2055}{2128}}{2(0.9)}\right)^2 - \frac{\frac{2055}{2128}}{0.9}} = 0.75$$

$$C_{px} = \frac{1 + \frac{1952}{2128}}{2(0.9)} - \sqrt{\left(\frac{1 + \frac{1952}{2128}}{2(0.9)}\right)^2 - \frac{\frac{1952}{2128}}{0.9}} = 0.73$$

C_P is the lesser of C_{Px} and C_{Py}, therefore use $C_P = 0.73$ and $F'_c = F*_cC_P = 2128(0.73) = 1553$ psi

The answer is **(A)**

Solution S28

To resist tension, determine the withdrawal per inch of a single nail

$W' = WC_DC_M{}^2C_tC_{eg}C_{tn}$

C_D = 1.0 since the controlling load is DL + LL
C_M = 1.0 since Moisture content is < 19%
C_t = 1.0 under normal temperature
C_{eg} = 1.0 since connection is not at the end grain
C_{tn} = 1.0 connection is not a toe nail

Therefore W = W'

$W = 1380G^{5/2}D$ for nails, Determine G from Table 12.3.3A for Red Pine G = 0.44

$W = 1380(0.44)^{5/2}(0.177)$ = 31.37 lb/in(2" embedment) = 62.74 lbs/nail

Number of nails = 500 lbs/62.74 lbs/nail = 7.97, use 8 nails,

The answer is **(C)**

Solution S29

The resultant $R = \sqrt{(V_P + V_e)^2 + H_e{}^2}$

$V_p = P/N = 50/8 = 6.25$ K, $V_e = Pex_c/I_o$, $H_e = Pey_c/I_o$

The moment of inertia, I_o, is found as a per area of bolt since all bolts have the same diameter. Simply sum the distances of each bolt to the appropriate neutral axis

$I_o = \Sigma y^2 + \Sigma x^2 = 4(1)^2 + 4(3)^2 + 8(4)^2 = 168$ in^4/in^2

$V_e = 50(12)(4)/168 = 14.29$, $H_e = 50(12)(3)/168 = 10.71$

$R = \sqrt{(6.25 + 14.29)^2 + 10.71^2}$ = 23.16 K The answer is **(A)**

Solution S30

The first number in the label of a wooden member indicates the reference allowable bending stress. The answer is **(C)**

Solution S31

It is important to remember conceptually that the purpose of prestressed or post-tensioned cables is to offset the tension from loading conditions. Therefore, the strand profile should be placed in the tension zone throughout the member. The answer is **(A)**

Solution S32

To determine the stress first identify all the contributing forces. There are 3 components to the total stress:

-Bending stress due to loading: Mc/I
-Compression stress due to the strands: P/A
-Bending due to the eccentricity of the strands: Pe(c)/I

Determine the Moment of Inertia of the beam $I = 1/12(8)(10)^3 = 666.67\ in^4$

Determine the max moment due to the distributed load $= wl^2/8 = 2(50)^2(12in/ft)/8 = 7500$ k-in

Be very careful of signs positive is tension and negative is compression

Total Bending Stress = Mc/I – P/A – Pe(c)/I = 7500(5)/666.67 – 100/(10(8)) – 100(3)(5)/666.67 = 52.75 ksi

The answer is **(B)**

Solution S33

PCI provides provision for the handling of precast members. To ensure no discernible cracking the modulus of rupture of the section must be greater than stress due to the applied moment during handling. First determine the applied moment from the equations provided for a 4-point pick:

$M = 0.0107wab^2$, w = 0.15 pcf(0.5') = 0.075 ksf, $M = 0.0107(0.075)(6)(40)^2 = 7.7$ K – ft(12) = 92.4 K-in

Determine the moment of inertia $= 1/12(72)6^3 = 1296\ in^4$

Applied stress with the 1.2 multiplier for yard handling = 1.2Mc/I = 92.4(3)(1.2)/1296 = 0.256 ksi = 256 psi

The modulus of rupture is essentially the tensile strength of the concrete

$F_r = 5\sqrt{f'_c} = 256$. Solve for $f'_c = 2621$ psi. The answer is **(A)**

Solution S34

The thermal range is the elongation or contraction which can occur between the high and low temperatures respectively. This question only asks for the contraction. For steel, thermal expansion or contraction is determined by the following equation:

$\Delta = L_o\alpha(t_1 - t_o)$

To determine the contraction, compare the low temperature to what the bridge is set at originally

Δ = 120(0.0000065)(10-65)(12) = -0.5" The answer is **(B)**

Solution S35

The center of gravity can only be determined for a single material. Therefore, the area of steel needs to be converted into an equivalent area of concrete. This is done using the ratio of moduli of elasticity $n = E_s/E_c$ = 29000/3605 = 8.044

Calculate the equivalent width of concrete for the steel b_{equiv} = bn = 8.04(8) = 64.36.

Then determine the center of gravity

	A	y	Ay
1	8(12) = 96	6.5	624
2	64.36(1/2) = 32.18	0.25	8.045
Sum	128.2		632

The center of gravity from the bottom = 632/128.2 = 4.9" The answer is **(A)**

Solution S36

The effective depth is essentially finding the center of gravity of the reinforcing accounting for the distribution of area. Add up the areas and distances from the bottom face:

	A	**Y**	**AY**
1	2(0.44)=0.88	6	5.28
2	3(0.79)=2.37	4	9.48
3	4(1.0)=4.0	2	8.0
SUM	7.25		22.76

22.76/7.25 = 3.14 Subtract this from the overall height 12 – 3.14 = 8.86" The answer is **(D)**

Solution S37

The guidelines for weld symbols can be found in AISC Table 8-2.1. The weld shown is a fillet weld both on arrow side and opposite side, therefore the appropriate symbol is the triangle on the top and bottom. A field weld is indicated by a flag. The answer is the following:

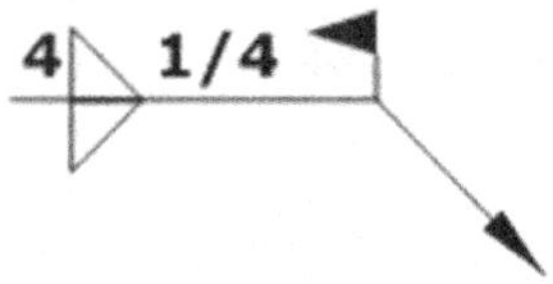
4
1/4

Solution S38

Limiting steel to a maximum ensures the failure mode is not crushing of the concrete and thus will fail in a non-brittle process. The answer is **(C)**

Solution S39

AWS Section 2.18 describes prohibited weld types. The answer is **(A)**

Solution S40

Horizontal distance requirements for safety nets are found on the table in Sec. 1926.502-c2. For >10’ drop off the net needs to be 13’. The answer is **(D)**

Solution S41

The governing concept is to understand the effect of loading conditions on continuous spans. A load on one span will have the opposite effect on adjacent spans and similar effects 2 spans away. Therefore, the answer is:

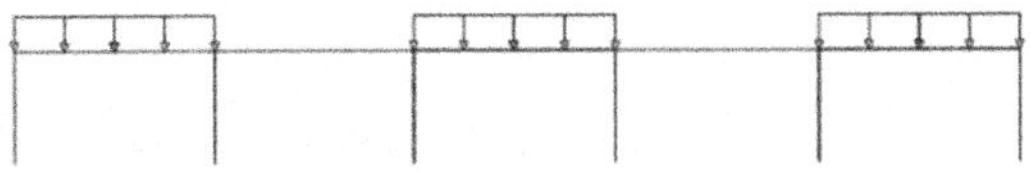

Solution S42

As per OSHA section 1910.23-e1, a standard railing must be 42" in height. The answer is **(B)**

Solution S43

First determine the torsional moment from the wind load T = 5(10)(12 in/ft) = 600 K – in

The torsional shear stress $\tau = Tr/J$, for a hollow circular member $J = \pi/2(r_o^4 - r_i^4) = \pi/2((6)^4 - (4.5)^4) = 1391.63\ in^4$

$\tau = 600(6)/1391.63 = 2.6$ ksi The answer is **(A)**

Solution S44

Horizontal shear stress $\tau = VQ/Ib$

$Q = ay = (3)(8)(7.5 - 1.5) = 144\ in^3$, $I = 1/12bh^3 = 1/12(8)(15)^3 = 2250\ in^4$

$\tau = 50(144)/(2250(8)) = 0.4$ ksi = 400 psi, The answer is **(A)**

Solution S45

From table 19.3.1.1, the exposure class can be determined. The descriptions yield W1, C2, and F1. Then from table 19.3.2.1 the minimum values can be matched as follows:

	w/c	f'c (psi)
Concrete in contact with water and low permeability is required	0.50	4000
Concrete exposed to moisture and an external source of chlorides	0.40	5000
Concrete exposed to freezing and thawing cycles with limited exposure to water	0.55	3500

Solution S46

First using the ground snow load, determine the density of the snow:

$\gamma = 0.13p_g + 14 < 30$ pcf $= 0.13(40) + 14 = 19.2$ pcf

Then using the density determine the height of the roof snow $h_b = p_s/\gamma = 25/19.2 = 1.3'$

Therefore $h_c = 10 - 1.3 = 8.7'$

Then using figure 7-9 determine the drift height h_d
For Leeward drift use $l_u = 200'$, use approx. $h_d = 4.75'$
For Windward use $l_u = 600'$, which gives a value approx. 7.75. Use $h_d = 3/4(7.75) = 5.8'$

Use the larger of the 2 values, therefore $h_d = 5.8'$

Now find w, since $h_d < h_c$ $w = 4h_d = 4(5.8') = 23.25'$

To find the load/ft, find the area of the triangular snow load and multiply by the density $\frac{1}{2}(w)(h_d)\gamma = \frac{1}{2}(23.25)(5.8)(19.2) = 1294$ lb/ft The answer is **(B)**

Solution S47

The effective weight is the load which can be accounted for to offset horizontal seismic loading. This includes the dead load and any additional loading as outlined in the 5 rules in Section 12.7.2. Rules 1 and 4 apply here since the first floor is used for storage and the Snow Load is greater than 30. Therefore:

W = 200(30 + 0.2(35) + 25 + 0.25(40)) + 5000 = 19400 lbs = 19.4 K. The answer is **(A)**

Solution S48

Considering the building is a seismic category A, there actually is no seismic force requirements as per section 11.7. However, it is still required to design for a notional lateral force as per section 1.4.

$F_x = 0.01W_x = 0.01(150) = 1.5$ K The answer is **(B)**

Solution S49

As per table 1808.8.1 the minimum strength of a micro pile is 4000 psi.

Solution S50

The IBC has stricter requirements on concrete cover for certain foundation elements. As per table 1808.8.2 a precast prestressed deep foundation element exposed to seawater has a minimum cover of 2.5". The answer is **(B)**

Solution S51

Each limit state is defined in AASHTO Chapter 3. Service II is used for the yielding of steel. The answer is **(C)**

Solution S52

Need to position the truck where the shear will be greatest. This will be putting the heavier axle at the support. Then take moments about the opposite support to determine the reaction:

0 = V40 – 14(40) – 8(32), V = 20.4 K

The answer is **(C)**

Solution S53

Use the provisions of ACI section 22.6.5.2. First determine the perimeter of the critical section. This is determined by creating a rectangle a distance of d/2 from each face of the column.

$b_o = 2(36 + 2(12/2)) + (48 + 2(12/2))2 = 216''$

The 2-way shear strength is the smallest of the following 3 equations:

$(2+4/\beta)\sqrt{f'_c}b_od$ Where β = the ratio of the long side to the short side of the column

$= (2+ 4/(4/3))\sqrt{4000}/1000(216)(12) = 819.5$ K

$(\alpha_s d/b_o+2)\sqrt{f'_c}b_od$ Where $\alpha_s = 40$ for interior columns

$= ((40(12))/216+2)\sqrt{4000}/1000(216)(12) = 691.6$ K

$4\sqrt{f'_c}b_od = (4\sqrt{4000})/1000(216)(12) = 655.9$ K

$\phi V_n = 0.75(655.9) = 491.7$ K The answer is **(A)**

Solution S54

Determine the appropriate load factors for each moment. The trick is you will notice for DC and DW factors there is a maximum and minimum load factor. AASHTO requires that force effects which produce an opposite effect will be factored by the minimum load factor. Therefore, the factored moment is:

1.25(25) + 0.90(-5) + 1.5(10) + 0.65(-2) + 1.75(50) = 128 K-ft, The answer is **(B)**

Solution S55

From section ACI 25.7.2 use the minimum of the following:

1. 16 Longitudinal bar diameters = 16(1.0") = 16"
2. 48 tie bar diameters = 48(0.5") = 24"
3. Least dimension of the column = 36"

The answer is **(A)**

Solution S56

The capacity of steel compression members is a function of the unbraced length. However, since W shapes are not doubly symmetric, there is a different capacity about each axis and it must be determined which is the controlling axis. To do this AISC Table 4-1 provides the conversion factor to determine the equivalent unbraced length of the stronger axis which is the ratio of the radii of gyration about each axis. The unbraced length about the X-X axis is given as 38. Since the column is braced at 20', the unbraced length about the Y-Y = 20'. Then determine the equivalent unbraced length for the X-X axis:

$(KL)_{Yequiv} = (KL)_X/(r_x/r_y) = 38/1.59 = 23.9 > 20'$ Therefore the critical unbraced length is the larger of these values which is 24'.

Therefore, from the Table 4-1, the design capacity is 1310 Kips

The answer is **(B)**

Solution S57

A steel tension member needs to be checked for 2 modes of failure. Yielding of the gross area and rupture of the effective net area.

$\phi R_n = F_y A_g$ Yielding $\phi = 0.9$
$\phi R_n = F_u A_n$ Rupture $\phi = 0.75$

The gross area includes no holes $A_g = 10t$
The net area includes the holes. The diameter of the hole is obtained by adding 1/8" to the diameter of the bolt $A_n = t(10-(7/8 + 1/8)3) = 7t$

Solve for t

Yielding $100 = (0.9)36(10t)$, $t=0.3$
Rupture $100 = (0.75)58(7t)$, $t=0.328$

The answer is **(B)**

Solution S58

First determine the maximum applied moment by adding the effects from the concentrated load and distributed load

$M = wl^2/8 + PL/4 = 2(20)^2/8 + 10(20)/4 = 150$ k-ft

There may be a tendency to apply the appropriate equations however it is more efficient to utilize the design table 3-10 in AISC. Simply find the applied moment and match with the unbraced length. The most efficient section is the first dark line above the point where the 2 intersect.

The answer is W14x43. The answer is **(C)**

Solution S59

Fatigue is covered in the specifications Appendix 3 of the AISC. The stress range is determined by the following equation:

$F_{SR} = (C_f/N)^{0.333} > F_{TH}$

N is the fluctuations for the life of the structure = 100 per day(365 days/year)(50 years) = 1.825×10^6

C_f = Found from table A3.1 = 120×10^8

F_{TH} = Also found on the same table = 16 ksi

$F_{SR} = (120 \text{ X } 10^8/ 1.825 \times 10^6)^{0.333} = 18.67$ ksi > 16 The answer is **(C)**

Solution S60

The design bending stress is determined by applying all appropriate modification factors to the reference bending stress

$F_b' = F_bC_DC_mC_tC_LC_VC_{fu}C_CC_I$ however only use the lesser of C_L or C_V
F_b = 2400 psi as given from the designation 24F
C_D = 1.15 since the controlling load case includes snow load
C_M = 0.8, moisture content is >16%
C_t = 1.0 under normal temperature conditions
C_L = 1.0 since the beam is fully laterally supported
$C_V = (21/L)^{1/x}(12/d)^{1/x}(5.125/b)^{1/x} = (21/20)^{1/20}(12/6)^{1/20}(5.125/6)^{1/20} = 1.03$
C_{fu} = 1.0 for bending about the strong axis
C_C = 1.0 for no curvature

C_l = 1.0 since there is no taper

F_b' =2400(0.8)(1.15)(1.0)(1.0)(1.0)(1.0)(1.0) = 2208 psi The answer is **(B)**

Solution S61

For a wood beam design in shear as per section 3.4, the distributed load is neglected from each end of the beam a distance of the support width + d. Therefore:

V = w(L – 2(b+d))/2 = 0.5(20 – 2(1 +30/12))/2 = 3.25 K The answer is **(A)**

Solution S62

The service moment capacity is the lesser of the following 2 equations:

$M_m = F_b jkb_w d^2/2$.   Or $M_S = F_s j\rho b_w d^2$

F_s = 32 ksi for grade 60 steel, $F_b = f'_m 0.45$ = 1.5(0.45) = 0.675 ksi, $\rho = A_s/b_w d$ = 1.0/(6.5(36)) = 0.0043

$n = E_s/E_m = 29000000/900f'_m$ = 29000000/900(1500) = 21.48

$$k = \sqrt{(2n\rho + pn^2)} - \rho n$$
$$= \sqrt{(2(0.0043)(21.48) + 21.48n^2)} - (0.0043)(21.48) = 0.347$$

J = 1 – k/3 = 1 – 0.347/3 = 0.884

M_m = 0.675(0.884)(0.347)(6.5)(36)^2/(2(12 in/ft)) = 72.67 ft – K

M_S = 32(0.884)(0.0043)(6.5)(36)^2/(12 in/ft) = 85.4 ft – K

Therefore use 72.67 ft – K. The answer is **(A)**

Solution S63

First determine the appropriate equation by calculating the slenderness ratio = h/r

$r = 0.289b$ for squares = 0.289(14) = 4.05 in. $h/r = 20(12)/4.05 = 59.26 < 99$ Therefore use equation 2-20

$P_a = (0.25f'_cA_n + 0.65A_{st}F_s)(1 - (h/(140r)^2) = (0.25(2)(14(14) - 1.76) + 0.65(1.76)(32))(1 - (20(12)/(140(4.05)))^2) = 109$ K

The answer is **(A)**

Solution S64

The requirements for shear reinforcement are found in MSJC 8.1.6.6 and 8.3.5.2. The reinforcement does not need to create a closed loop. The answer is **(C)**

Solution S65

Options A, C, and D are all reinforcement requirements as per MSJC section 6.1. The reinforcement strength is not limited to 65 ksi. The answer is **(B)**

Solution S66

Contributing forces to the moment about the X-X axis only are the eccentric force of the sign and the wind on the sign

Sign Dead Load = 1(2') = 2 ft – K
Wind = 0.015 ksf(5')(5')(20-1) = 7.125 ft – K

However, the moments are acting in opposite directions

$M_{X\text{-}X} = 7.125 - 2 = 5.125$ K – ft The answer is **(A)**

Solution S67

First find the reaction at B by taking moments about A

$\Sigma M_A = 0 = 10(20) + 5(30) + 2(40) - 10B$, B = 43 K

Find the vertical and horizontal reaction at A

Sum of the vertical forces = 0 = 10 + 5 + 2 – A_V, A_V = 17 K

Sum horizontal forces = 0 = 43 – A_H, A_H = 43 K

The free body diagram of the truss is shown below:

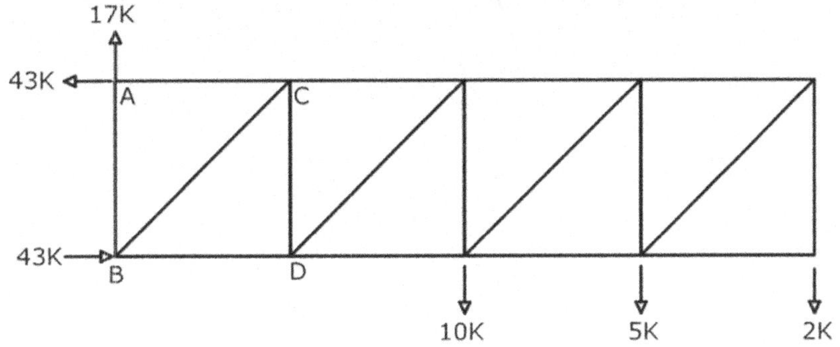
17K
43K
43K
A
B
C
D
10K
5K
2K

Next taking a free body diagram at point A and summing vertical forces indicates Member AB = 17 K

Then taking a free body diagram at point B, the force in member BC can be determined.

Sum vertical forces = 0 = 17 – BCsin(45), BC = 24.0 K

Then summing horizontal forces = 0 = 43 – 24cos(45) – BD, BD = 26 K The answer is **(C)**

Solution S68

As per AASHTO, DW refers to the wearing surface on the bridge. This has a different load factor due to the anticipation of the thickness to wear away. The answer is **(A)**

Solution S69

AISC 3-221 provides equations for continuous beams uniformly loaded in various situations. For this question the max moment is:

$M=0.101wl^2 = 0.101(2)(20)^2 = 80.8$ K-ft The answer is **(A)**

Solution S70

To analyze a hinged beam, remember that the moment is 0 at that point. Therefore, if you analyze the beam by taking a free body diagram to the left of the hinge, you can determine the vertical reaction at the hinge by taking moments about A:

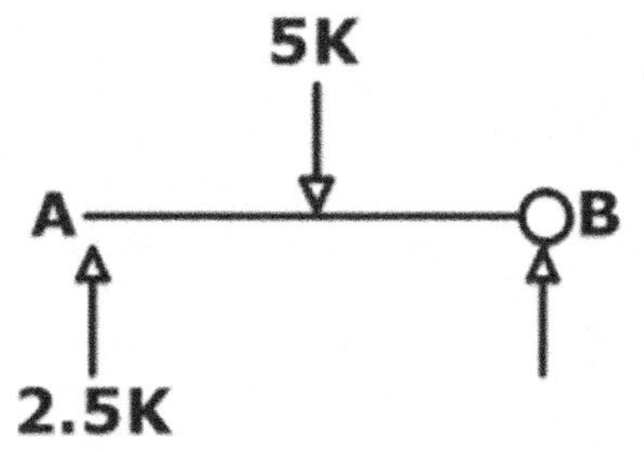
5K
A
B
2.5K

B = 5(2.5)/5 = 2.5 K

Then taking a free body diagram to the right of the hinge, determine the reaction at C

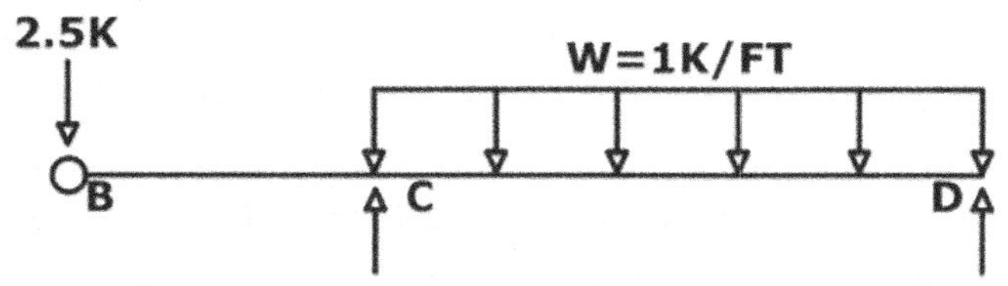
2.5K
W=1K/FT
B
C
D

summing moments about D = 0 = -2.5(15) + 10C – 1(10)(5), C = 8.75 K The answer is **(D)**

Solution S71

The important concept is to be able to convert a moment into a force couple. Following the load path, the force creates a moment on the slab:

M = 10(20) = 200 ft-kips.

This moment is then resisted by a force couple in the piles. Simply divide the moment by the distance

P = 200/30 = 6.67 K The answer is **(A)**

Solution S72

Divide the force into its components:

$F_Z = 0$

$F_X = 5\cos(32) = 4.24$ K

$F_Y = 5\sin(32) = 2.63$ K

Determine moments:

$M_X = 2.65(10) = 26.5$ K-ft

$M_Y = 4.24(10) = 42.4$ k-ft

$M_Z = 4.24(6) + 2.65(10) = 51.54$ K-ft

The answer is **(C)**

Solution S73

IBC Table 1705.3 shows verification and inspection requirements and identifies which types require continuous vs periodic inspection. The ones that require continuous include slump and air content tests, shotcrete placement for proper application techniques, and inspection of prestressed concrete. The answer is **B, E, and F**

Solution S74

Allowable pressure without an applied moment $q_a = P/A_f + \gamma_c h + \gamma_s(H\text{-}h)$, where $A_f = L_w$. Therefore, to find the length solve for L:

$2000 = 40000/6L + 150(3) + 75(2)$ solve for L = 4.8′ The answer is **(A)**

Solution S75

First take free body diagram of the system to determine Reactions:

$\Sigma M_A = 3(10) + 5(14) - D_Y(20) - D_X(2)$

However, we have one equation and 2 unknowns. We need to relate the reactions at D to each other. We can take a free body diagram of point D

$\Sigma F_X = D_X - DC\cos(18)$ Therefore $D_X = DC\cos(18)$

$\Sigma F_y = D_y - DC\sin(18)$ Therefore $D_y = DC\sin(18)$

Now you can find DC by plugging into the moment equation above:

$\Sigma M_A = 3(10) + 5(14) - DC\sin(18)(20) - DC\cos(18)(2)$, DC = 12.4 K

Now take a free body diagram of point C and sum the horizontal forces to find tension in BC:

$\Sigma F_X = 12.4\cos(18) - BC\cos(8)$

BC = 11.9 K

The answer is **(B)**

Solution S76

First determine the maximum moment due to the loading conditions. For a simply supported beam with 2 point loads equal distance from the ends, the equation for maximum moment is:

M = Pa = 10(10) = 100 ft–K(12in/ft) = 1200 in - K

The stress is determined by Mc/I, $I = 1/12(8)(12)^3 = 1152\ in^4$

The depth of the desired stress is not at the extreme fiber. Therefore c = 6 - 3 = 3″
Stress along A-B = 1200(3)/1152 = 3.125 ksi The answer is **(B)**

Solution S77

The HL-93 loading was established as the loading criteria for LRFD. It is a combination of the HS-20 truck as well as an additional distributed Live Load. The answer is **(C)**

Solution S78

The column end conditions is fixed-pinned and therefore K = 0.8. However, the braces reduce the length to segments with the longest being 50′. The effective length factor also needs to be applied to this segment and KL = 0.8(50) = 40′ The answer is **(B)**

Solution S79

First determine the moment of inertia of each beam and the modulus of elasticity

$I_1 = (1/12)(12)18^3 = 5832\ in^4$, $I_2 = (1/12)(8)(12)^3 = 1152\ in^4$

E = 57000$\sqrt{f'_c}$ = 57000$\sqrt{4000}$/1000 = 3605 ksi

Then draw the moment diagram. Since this is a cantilever beam with only a moment couple applied, the moment diagram is constant:

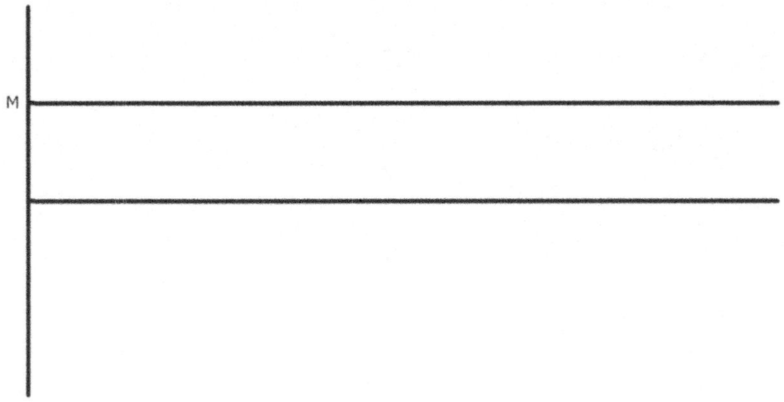
M

Then construct the M/EI diagram by dividing the moment at each point by the EI at that point. Since the beam is not a uniform cross section, this diagram looks as follows:

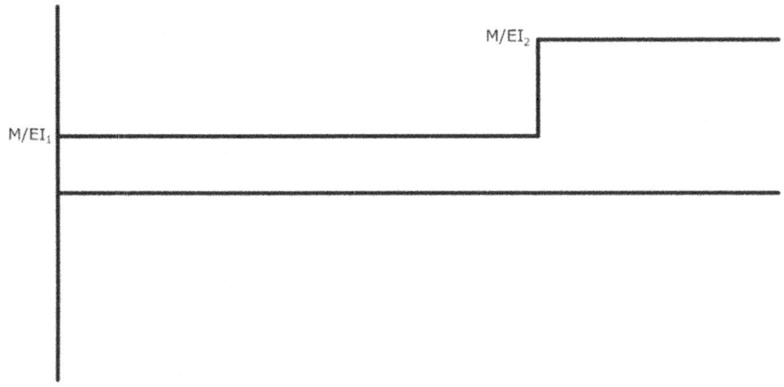
M/EI₂
M/EI₁

To calculate the deflection, calculate the area under the M/EI diagram and multiply this by the distance from the centroid of the area to the free end. The equation comes to the following:

$$\frac{M}{EI_1}(20')\left(10'+\frac{20'}{2}\right)+\frac{M}{EI_2}(10')\left(\frac{10'}{2}\right)$$
$$=\frac{10(12)}{(3605)(5832)}(20'(12))\left(10'(12)+\frac{20'(12)}{2}\right)+\frac{(10)12}{3605(1152)}(10'(12))\left(\frac{10'(12)}{2}\right)$$
$$=0.329+0.208=0.54"$$

The answer is **(C)**

Solution S80

The elongation of a beam due to a tension load is determined from the following equation:

$\Delta = PL/AE$

The properties of the W14x100 beam can be found in AISC, $A=32.0\ in^2$

$\Delta = 100(50)(12)/32(29000) = 0.065''$ The answer is **(D)**

Answer Key

M1	A	S1	C	S41	See Solution
M2	C	S2	B	S42	B
M3	ACGDH	S3	B	S43	A
M4	D	S4	C	S44	A
M5	A	S5	A	S45	See Solution
M6	B	S6	D	S46	B
M7	See Solution	S7	B	S47	A
M8	A	S8	A	S48	B
M9	C	S9	A	S49	4000
M10	D	S10	D	S50	B
M11	B	S11	D	S51	C
M12	B	S12	A	S52	C
M13	10+60	S13	C	S53	A
M14	A	S14	B	S54	B
M15	D	S15	B	S55	A
M16	A	S16	D	S56	B
M17	C	S17	A	S57	B
M18	D	S18	A	S58	C
M19	B	S19	A	S59	C
M20	A	S20	See Solution	S60	B
M21	C	S21	ABCEF	S61	A
M22	A	S22	A	S62	A
M23	D	S23	A	S63	A
M24	C	S24	B	S64	C
M25	A	S25	C	S65	B
M26	See Solution	S26	D	S66	A
M27	B	S27	A	S67	C
M28	C	S28	C	S68	A
M29	A	S29	A	S69	A
M30	A	S30	C	S70	D
M31	B	S31	A	S71	A
M32	D	S32	B	S72	C
M33	D	S33	A	S73	BEF
M34	C	S34	B	S74	A
M35	A	S35	A	S75	B
M36	C	S36	D	S76	B
M37	C	S37	See Solution	S77	C
M38	C	S38	C	S78	B
M39	B	S39	A	S79	C
M40	B	S40	D	S80	D

Thank You Again for Your Purchase!

For an additional 40 Civil Practice Problems enter your email address at the following web address:

www.PECoreConcepts.com/subscribe

References

AASHTO LRFD Bridge Design Specifications, 7th edition,
American Association of State Highway & Transportation Officials,

IBC International Building Code, 2015 edition, International Code Council.

ASCE 7 Minimum Design Loads for Buildings and Other Structures, 3rd printing, 2010, American Society of Civil Engineers.

ACI 318 Building Code Requirements for Structural Concrete, 2014, American Concrete Institute.

AISC Steel Construction Manual, 14th edition, American Institute of Steel Construction, Inc.

NDS National Design Specification for Wood Construction, 2015 edition, and National Design Specification Supplement, Design Values for Wood Construction, 2015 edition, American Wood Council.

OSHA CFR 29 General Industry regulations and Construction regulations, 2016

PCI Design Handbook: Precast and Prestressed Concrete, 7th edition, 2010, Precast/Prestressed Concrete Institute.

TMS 402/6023 (ACI 530/530.1) Building Code Requirements and Specifications for Masonry Structures, 2013; The Masonry Society.

Civil Engineering Reference Manual for the PE Exam Fifteenth edition 2015. Michael R. Lindeburg Professional Publications Inc. (PPI)

Made in the USA
Las Vegas, NV
08 December 2023

82275088R00105